Bluetooth 5.0 Modem Design for IoT Devices

Khaled Salah Mohamed

Bluetooth 5.0 Modem Design for IoT Devices

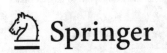 Springer

Khaled Salah Mohamed
Siemens Digital Industries Software
Fremont, CA, USA

ISBN 978-3-030-88628-8   ISBN 978-3-030-88626-4   (eBook)
https://doi.org/10.1007/978-3-030-88626-4

This Springer imprint is published by the registered company Springer Nature Switzerland AG
The registered company address is: Gewerbestrasse 11, 6330 Cham, Switzerland

To my beloved mother, Prof. Layla Kotb, who raised me to be the person I am today. Thank you for all the unconditional love that you have given me, encouraging me to achieve all my dreams. Thank you for everything. May god give you an eternal life.

Preface

Among various beacon schemes, Bluetooth Low Energy (BLE) beacon is one of the most promising systems for enabling IoT. Bluetooth Low Energy (BLE) is widely used as a low-power communication protocol for short-range IoT sensors. As the BLE standard requires greatly relaxed specifications compared to the classic Bluetooth standard, the implementation of a low-power BLE transceiver with a simple structure is possible.

Bluetooth Low Energy (BLE) is considered as a short-range, energy-efficient, low-power, and low-cost radio technology for enabling IoT. Bluetooth is short-range wireless communication standard designed with an intention of replacing cables connecting portable and desktop devices and building wireless networks for such devices and enabling communication to be established between these devices up to maximum distance of 100 meters without obstacles. Bluetooth features robustness, low cost, low complexity, and low power; it uses the free industrial, scientific, and medical (ISM) frequency band (2.4–2.48 GHz).

The Bluetooth specification is an open specification that is governed by the Bluetooth Special Interest Group (SIG). The Bluetooth SIG is led by its five founding companies and four new member companies who were added in late 1999.

This book provides an introduction to Bluetooth technology, with a specific focus on developing a hardware architecture for its modem. The major concepts and techniques involved in Bluetooth technology are discussed, with special emphasis on hardware mapping. The book starts simply to allow the reader to quickly master the basic concepts before addressing the advanced features.

This book differs from the researches published up to date in that it presents Bluetooth transceiver architecture suitable for implementation in an FPGA. It examines several digital algorithms for modulation and demodulation of Bluetooth signals, locking on the carrier phase, and synchronizing the symbol. Many of these previously analog designs have been translated to the digital domain.

The existing literature does not study modem architectures to determine an effective implementation to handle the variation in bit rate due to different constellations while minimizing redundant hardware required to operate on multiple constellations in an FPGA. Literature is also deficient on synchronization techniques for multiple constellations and the implementation of these techniques in an FPGA.

Unlike ASICs, FPGAs are reconfigurable, that is, their internal structure is only partially fixed at fabrication, leaving to the application designer the wiring of the internal logic for the intended task. This can significantly shorten design and production, and thus time to market, for FPGA-based systems. Although FPGAs tend to be slower and consume more power than ASICs, FPGA re-configurability can benefit platform longevity (which is extremely important in an era of fast-changing wireless communications standards) by allowing design changes/upgrades even in systems already in operation. This flexibility can be effectively exploited for rapid prototyping of advanced communications signal processing.

The main design issues for Bluetooth transceivers are not only low cost and low power consumption, but also quality performance. Classical designs of the Bluetooth receiver utilize data-aided techniques to correct carrier frequency offsets and symbol timing errors. Such techniques offer low-cost and reasonable performance. Non-data-aided techniques offer an alternate higher-performance approach to correct the same problems, at the penalty of an increased hardware complexity and cost.

Another purpose of this book is to investigate the trade-off between cost and performance when a Bluetooth transceiver is designed using non-data-aided techniques for clock and timing recovery. The Bluetooth transceiver supports (GFSK, π/4DQPSK, 8DPSK) modulation techniques and is able to achieve data rates of 1 Mbps, 2 Mbps, and 3 Mbps.

Our proposed Bluetooth transceiver design is encoded in the VHDL hardware description language and implemented successfully on Spartan 3 Xilinx FPGA. The performance of the transceiver is experimentally tested using the Xilinx FPGA-embedded *ChipScope* logic analyzer. Non-data-aided timing recovery improves the BER performance by as high as 3 dB at an area penalty of ~15% of the total design size. Verifications at worst-case frequency offset show an error rate performance of 10^{-3} at E_b/N_0 of 15 dB for GFSK, 10^{-4} at E_b/N_0 of 13.4 dB for π/4DQPSK, and 10^{-4} at E_b/N_0 of 20 dB for 8DPSK.

Fremont, CA, USA   Khaled Salah Mohamed

Contents

About the Author

Khaled Salah Mohamed received his B.Sc. degree in Electronics and Communications Engineering with distinction and honor degree in 2003 from Ain-Shams University, Cairo, Egypt. He received his M.Sc. and his Ph.D. degrees in electronics and communications in 2008 and 2012, respectively. Dr. Salah received his MBA degree in 2016. He joined Mentor Graphic Corporation, where he designed many SoC IPs such as AHB, HDMI, HDCP, eMMC, SDcard, HMC, and LPDDR5. Moreover, Dr. Salah worked as an engineering lead at the emulation division at Mentor Graphic, Egypt. Currently, he is an applications engineering consultant at Mentor Graphic, USA. Dr. Salah has published 5 books and more than 100 research papers in the top refereed journals and conferences. His research interests are in 3D integration, IP modeling, Internet of Things, artificial intelligence, machine learning, and SoC design. He is a senior IEEE member. Dr. Salah served as a reviewer for several conferences and journals, including *IEEE Transactions on VLSI, IEEE Transactions on Circuits and Systems II, IEEE Transactions on Semiconductor Manufacturing, IEEE Microwave and Wireless Components Letters, IEEE Transactions on Microwave Theory and Techniques*, and *Elsevier's Microelectronics Journal*.

Chapter 1
An Introduction to Bluetooth

1.1 Bluetooth History

A Bluetooth device uses radio waves instead of wires or cables to connect to a phone or computer. A Bluetooth product, like a headset or watch, contains a tiny computer chip with a Bluetooth radio and software that makes it easy to connect. When two Bluetooth devices want to talk to each other, they need to pair. Communication between Bluetooth devices happens over short-range, ad hoc networks known as piconets. A piconet is a network of devices connected using Bluetooth technology [1]. When a network is established, one device takes the role of the master while all the other devices act as slaves. Piconets are established dynamically and automatically as Bluetooth devices enter and leave radio proximity [2–4]. Piconet is basic structure of Bluetooth communications that consists of a master device and at most 7 slave devices. Usually, a device with the least computation power is selected as master device and the others are selected as slave devices [5, 6].

One of the disadvantages of the original version of Bluetooth in some applications was that the data rate was not sufficiently high, especially when compared to other wireless technologies such as 802.11. In November 2004, a new version of Bluetooth, known as Bluetooth 2.0 was ratified. This not only gives an enhanced data rate but also offers other improvements as well [7, 8].

Bluetooth 5.0 is the latest version of the Bluetooth wireless communication standard. It is fully optimized for IoT [9–13]. It's commonly used for wireless headphones and other audio hardware, as well as wireless keyboards, mice, and game controllers. Bluetooth is also used for communication between various smart home and Internet of Things (IoT) devices. Bluetooth 5.0 offers twice the data transfer speed of the previous version, Bluetooth 4.2 while increasing the capacity of data broadcasts by 800% [14, 15]. With Bluetooth 5.0 you can send and receive much more data much more quickly. The new standard is effective over four times the range of Bluetooth 4.2, which will come in particularly useful for things like portable speakers. In January 2019, Bluetooth Special Interest Group (SIG) released

K. S. Mohamed, *Bluetooth 5.0 Modem Design for IoT Devices*,
https://doi.org/10.1007/978-3-030-88626-4_1

a new standard, Bluetooth 5.1 [16]. One of their new main features in this standard is the ability to detect the direction from which a Bluetooth signal is coming from, and claim they can estimate the direction with high accuracy. Previously standards have only had the feature for positioning by measuring the signal strength for distance. One method in version 5.1 is Angle of Arrival, it is a method that requires an antenna array with multiple antennas. The name of the Bluetooth standard originates from the Danish king Harald Blatand who was king of Denmark between 940 and 981 AD. His name translates as "Blue Tooth" and this was used as his nickname. A brave warrior, his main achievement was that of uniting Denmark under the banner of Christianity, and then uniting it with Norway that he had conquered. The Bluetooth standard was named after him because Bluetooth endeavors to unite personal computing and telecommunications devices [17–20].

1.2 Bluetooth Applications

Bluetooth has a wide range of application. The most common and expected applications will be described in this section. The Bluetooth security model includes the following processes namely pairing, passkey generation, device authentication, message verification. Bluetooth technology is considered as one of the most reliable and secure wireless connection. Bluetooth applications and the corresponding topology are shown in Table 1.1. There are billions of Bluetooth devices in use today. These devices are exposed to different types of threats. Bluetooth security solutions need to constantly evolve to mitigate emerging threats. Similar to any other wireless communication systems, Bluetooth transmissions can be deliberately jammed or intercepted. False or modified information can be passed to the devices by malicious users [21].

1.2.1 Handheld Devices

All information stored in a handheld device will be accessible from other Bluetooth enabled devices without connecting them by cable; of course it's also possible to exchange information from the other devices with a handheld. In this way it's possible to synchronize an agenda, phonebook, contact list, calendar and to-do list for all devices in a home or office environment. Connecting a PDA to a mobile

Table 1.1 Bluetooth applications and the corresponding topology

Application	Headsets	Internet access	Control systems Monitoring systems Automation systems
Topology	Point to point	Broadcast	Mesh

phone will enable somebody to synchronize all data even if one isn't in the office. Furthermore there is no worry that the file transfer will go wrong when you move on of your devices because there no requirement of line-of-sight.

1.2.2   Wire-Bound Connections

It's possible to connect a Bluetooth device to a wire-bound connection like: PTSN, ISDN, LAN or XDSL. This can be done with an access point.

1.2.3   Headsets

A Bluetooth headset enables someone to answer automatically his incoming calls, initiate a voice-activated dial-up and end a call on a mobile phone which of course is also Bluetooth enabled. But the headset can also be used on other Bluetooth enabled devices like a laptop. It even can be used to playback audio because a headset like this has very high sound quality.

1.2.4   Internet Access

Bluetooth is capable of connecting computers and handhelds with mobile phones and with wire-bound connections so connection to the Internet is very easy.

1.2.5   Localization

Localization has been recognized as one of the most important topics in wireless sensor networks (WSNs), where sensor nodes can be scattered randomly over a region and can get connected into a network. From the market for location services based on IoT, one can choose from the following main technologies: (1) Bluetooth Low Energy (BLE), (2) UWB (ultra-wide band), (3) Radio Frequency Identification. (RFID). When compared to other technologies, BLE provides a fully wireless hardware solution that is autonomous in terms of power source [22, 23].

1.3   Bluetooth Cores and Layers

There are many cores of Bluetooth; Bluetooth1.1 which is the basic data rate (1Mps) till Bluetooth 5.0 which has four rates. Table 1.2 provides an overview of the Bluetooth cores as shown below [1, 7]. The classic version of Bluetooth 5 is identical to the previous versions, while the significant innovations focus on the BLE version. It is clear that the new standard has been designed to create a communication network that ensures, over a short distance, a communication bandwidth that allows data exchange among the connected appliances and other smart devices of the IoT [24, 25]. Bluetooth low-energy (BLE) standard as a strong candidate to be the backbone of wireless IoT devices and support their low-power requirements [16, 26–33]. Table 1.3 shows an overview for the Bluetooth evolution.

The hardware level specification for the Bluetooth standard consists of four layers; the Radio layer, Baseband layer, Link Manager Protocol layer, and the Host Controller Interface layer [34, 35].

The Physical Layer is responsible for the electrical specification of the communication device, including modulation and channel coding. In the Bluetooth system this is covered by the Radio and part of the Baseband. The Radio is the lowest layer. Its interface specification defines the characteristics of the radio front end, frequency bands, channel arrangements, permissible transmit power levels, receiver sensitivity level, frequency hopping technique to avoid interference with other users in the ISM band from microwave oven to WI-FI as the transmission remains only in a given frequency for a short time, the standard uses a hopping rate of 1600 hops per second, these hops are spread over 79 fixed frequencies and they are chosen in a pseudo-random sequence, the fixed frequencies occur at 2400 + n MHz where the value of n varies from 1 to 79 where this gives frequencies of 2402, 2404 . . . . . 2480 MHz.

The next layer is the Baseband, which carries out Bluetooth's physical (PHY) and media access control (MAC) processing. This includes tasks such as device discovery, link formation, and synchronous and asynchronous communication with peer. Moreover, Bluetooth is unique in offering the front-end RF processing integrated with the baseband module. In this way, on-chip integration lowers the cost of the network interface and the small size makes it easy to embed Bluetooth chips in devices such as cell phones and PDAs.

Bluetooth peers must exchange several control messages for the purpose of configuring and managing the baseband connections. These message definitions are part of the link manager protocol (LMP).

Table 1.2   Bluetooth cores and transmission rates

Bluetooth core	Mode	Bit rate
1.1	Basic rate (BR)	1 Mbps
2.0	Enhanced rate (ER)	2 or 3 Mbps
5.0	Bluetooth low energy (BLE)	1 or 2 Mbps
		125 or 500 Kbps

Table 1.3 Bluetooth Evolution

Bluetooth specification	v1.1	v2.0 + EDR	v2.1 + EDR	v3.0 + HS	V4.0 + LE	v4.1	v4.2	v5.0
Year	2002	2004	2007	2009	2010	2013	2014	2016
Basic rate	YES	YES	YES	YES	YES	YES	YES	YESS
Enhanced date rate (EDR)	NO	YES	YES	YES	YES	YES	YES	YES
High speed (HS)	NO	NO	NO	YES	YES	YES	YES	YES
Low power (LE)	NO	NO	NO	NO	YES	YES	YES	YES

Table 1.4 Bluetooth modulation scheme

Specification Bluetooth core	Modulation	Bit rate
Bluetooth1.1	GFSK	1 Mbps
Bluetooth2.0	π/4 DQPSK	2 Mbps
	8DPSK	3 Mbps
Bluetooth5.0	GFSK	Four data rates

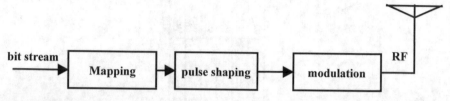

Fig. 1.1 Digital communication transmitter

The host controller interface (HCI) layer is implemented in Software and provides a command interface to the baseband controller and link manager, and access to hardware status and control registers.

Bluetooth data is transmitted as packets that have a standard format which consists of four elements:

(a) The Access Code which is used by the receiving device to recognize the incoming Transmission
(b) The Header which describes the packet type and its length.
(c) The Payload which is the data that is required to be carried.
(d) The Inter-Packet Guard Band which is required between transmissions to ensure that transmissions from two sources do not collide, and to enable the receiver to retune.

The important difference between the basic rate Bluetooth packet and the enhanced rate Bluetooth packet is that modulation scheme is changed within a high rate as basic rate specifications supports only binary modulation where the enhanced rate Bluetooth packet is M-ary modulated (π/4DPSK for 2 Mbps data rate, 8DPSK for 3 Mbps data rate) as shown in Table 1.4.

1.4 Transmitter Fundamentals

A simple block diagram of a digital communication transmitter is shown in Fig. 1.1 [36].

1.4.1   Mapping

For the first step in the modulation the digital data must be encoded into in-phase and quadrature phase (I,Q) signals. The following encoding schemes are used.

1.4.1.1   GFSK

Binary frequency shift keying (BFSK) is also known as Gaussian frequency shift keying (GFSK) in the terms of the Bluetooth application. No encoding schemes are used in BFSK modulation. The binary stream is used directly in the next step of modulation, in other words in the case of GFSK there are no Q signal. It is merely represented by the I signal which is the same as the bit sequence.

1.4.1.2 $\pi/4$ DQPSK

In $\pi/4$ DQPSK modulation technique every symbol represents two bits of the bit stream and therefore the bit rate is twice as large as the symbol rate. Every two bit is coded and now the bit stream is represented by I and Q signals as the I and Q signals now is calculated as:

$$I = Cos(\varphi) \qquad (1.1)$$

$$Q = Sin(\varphi) \qquad (1.2)$$

The difference between the QPSK and $\pi/4$ DQPSK is that in $\pi/4$ DQPSK the symbol is represented by the change of angle not by the angle itself, Table 1.5 shows the $\pi/4$ DQPSK mapping rules, detailed description of $\pi/4$ DQPSK modulation will be discussed in Chap. 3.

1.4.1.3   8DPSK

The same principles used in $\pi/4$ DQPSK are used for 8DPSK, but with 8DPSK three bits are transmitted with each symbol, meaning that the bit rate is three times larger than the symbol rate as shown in Table 1.6.

Table 1.5 $\pi/4$ DQPSK mapping rules

Bit 1	Bit 2	Δ angle
0	0	$\pi/4$
0	1	$3\pi/4$
1	0	$-3\pi/4$
1	1	$-\pi/4$

Table 1.6 8DPSK mapping rules

Bit 1	Bit 2	Bit 3	Δ angle
0	0	0	0
0	0	1	$\pi/4$
0	1	0	$\pi/2$
0	1	1	$3\pi/4$
1	0	0	π
1	0	1	$-3\pi/4$
1	1	0	$-\pi/2$
1	1	1	$-\pi/4$

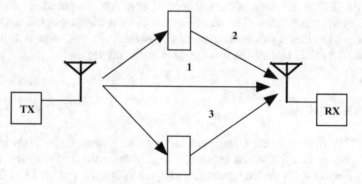

Fig. 1.2 Multi-path wireless channel

1.4.2   Pulse Shaping

1.4.2.1   The Concept of Inter-Symbol Interference (ISI)

As a signal travels in the channel, it is inevitably affected by various kinds of noise and fading in the channel. For example, in a wireless system whose channel is the free space between the transmitter and the receiver (Fig. 1.2), objects in propagation path can create multiple echoes on the transmitted signal. These echoes arrive at the receiver at different times and overlap to form a distorted version of the original signal. This effect of multiple echoes is referred to as multi-path fading.

In frequency domain, multi-path fading results in a non-uniform frequency response of channel, as some frequency components in the transmitted signal are enhanced and other frequency components are attenuated. One problem associated with a multi-path fading channel is inter-symbol interference (ISI), which refers to the interference to the current symbol by preceding symbols. Inter-symbol interference, which degrades the error-rate performance of the receiver, is a significant problem in data communication systems.

1.4.2.2   Pulse Shaping Techniques

When pulses are transmitted through a communications system, they are affected by the transmitter transfer function Ht(f), channel transfer function Hc(f), and receiver transfer function Hr(f) of that system. These three transfer function can be lumped together in one transfer function, H (f) where

$$H(f) = Ht(f) \ Hc(f) \ Hr(f) \tag{1.3}$$

The filtering with H(f) affects the received pulses in such a way that there is an overlap into adjacent symbol intervals.

These causes the pulses to change shape as the interfering pulses add constructively or destructively which may lead to errors in the detection process and this is called inter symbol interference (ISI).

Consider a system transmitting 1/T symbols/s, where T is the symbol time. The bandwidth required to detect 1/T symbols per second without ISI is equal to 1/2 T Hz, this condition is known as the Nyquist bandwidth constraint. This occurs when the system transfer function H (f) is made rectangular with bandwidth 1/2 T; this is as the ideal filter. The impulse response h(t) is the inverse Fourier transform of H (f) where

$$h(t) = \text{sinc}(t) \tag{1.4}$$

Such a pulse shape is not physical realizable due to the rectangular bandwidth characteristic, and the impulse response has in principle an infinite time delay so transfer functions which resembles the characteristics of the Nyquist filter and are realizable must be used in real systems.

Pulse shaping is performed with the following filters to avoid inter symbol interference (ISI): for GFSK a Gaussian filter is used (hence the name Gaussian Frequency Shift Keying), and for $\pi/4$ DQPSK and 8DPSK a square root-raised cosine filter is used. The Gaussian filter has an impulse response described by:

$$h(t) = \sqrt{\frac{\pi}{\alpha}} \exp\left(\frac{-\pi}{\alpha} t\right)^2 \tag{1.5}$$

$$\text{Where} \quad \alpha = \sqrt{\frac{\ln 2}{2\beta}} \tag{1.6}$$

Where β is the 3 dB bandwidth; Reducing the BT product improve the spectral efficiency as shown in Fig. 1.3.

The difference between GFSK and FSK is shown in Fig. 1.4 where GFSK is more spectral efficiency [37].

The square root-raised Cosine filter has a frequency response described by the following equation:

Fig. 1.3 The impulse response of the Gaussian pulse shaping filter

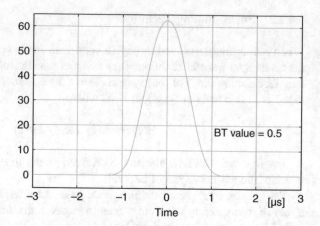

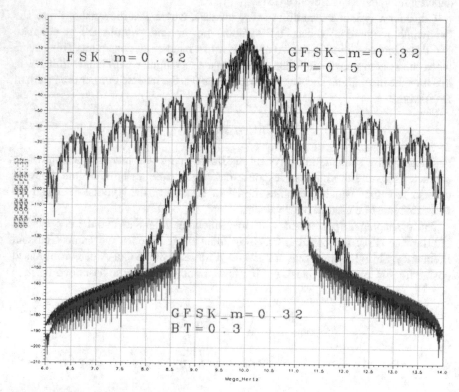

Fig. 1.4 GFSK spectral versus FSK

Fig. 1.5 Frequency
response of the square root-
raised cosine filter

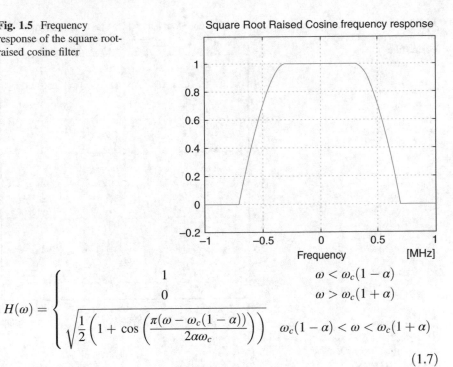

Square Root Raised Cosine frequency response

$$H(\omega) = \begin{cases} 1 & \omega < \omega_c(1 - \alpha) \\ 0 & \omega > \omega_c(1 + \alpha) \\ \sqrt{\dfrac{1}{2}\left(1 + \cos\left(\dfrac{\pi(\omega - \omega_c(1 - \alpha))}{2\alpha\omega_c}\right)\right)} & \omega_c(1 - \alpha) < \omega < \omega_c(1 + \alpha) \end{cases}$$

(1.7)

And it has impulse response h(t) described by the following equation:

$$h(t) = \frac{4\alpha}{\pi\sqrt{T}} \frac{\cos\left(\frac{(1+\alpha)\pi t}{T}\right) + \frac{T}{4\alpha \cdot t}\sin\left(\frac{(1-\alpha)\pi t}{T}\right)}{1 - \left(\frac{4\alpha \cdot t}{T}\right)^2}$$

(1.8)

Where $0 < \alpha < 1$. The frequency and impulse response of the square root-raised
cosine filter is shown in Figs. 1.5 and 1.6 respectively.

1.4.3 Modulation (Mixing with Carrier Frequency)

Mixing, or sometimes called waveform modulation, is the multiplication of a
sinusoidal wave and the signals I and Q.The frequency of the sinusoidal wave
corresponds to the modulation frequency, which henceforth will be referred to as
the intermediate frequency (IF).

Fig. 1.6 Impulse response
of the square root-raised
cosine filter

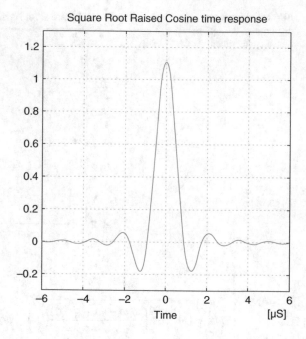

Square Root Raised Cosine time response

1.4.3.1   GFSK

The binary FSK I signal is mixed with two different modulation frequencies. The frequencies are chosen to be F1 = F_{if} + ∆F MHz and F2 = F_{if} − ∆F MHz, The GFSK modulation has bandwidth bit period product BT = 0.5. And the Modulation index shall be between 0.28 and 0.35 where the modulation index is given by:

$$h = \frac{1}{2} f_d T \tag{1.9}$$

And f_d is the frequency deviation, T the symbol time.

1.4.3.2   M-Ary DPSK

In the case of M-ary DPSK, the sinusoidal wave used for the Q part is 90 degree phase shifted from the I part, that is why I and Q-signals are called in-phase and quadrate signals respectively. The sinusoidal wave is multiplied with the I and Q signals, then The signals are added and can be written as:

$$S(t) = I \ \sin(2\pi f \ t) + Q \ \cos(2\pi f \ t) \tag{1.10}$$

1.5   Receiver Fundamentals

As with the transmitter in a digital communications system, the receiver will also be described in a simplified form. In Fig. 1.7, a block diagram of such a receiver is shown. In the outermost part of the receiver it consists of an antenna and RF mixers which is out of the scope of this Thesis then the demodulation stage [38].

1.5.1   Demodulation

The demodulation is the mixing with the same modulation frequencies as used in the transmitter part, after the mixing the signal is once again represented in I and Q signals. A more detailed description of demodulation is described in Chaps. 2 and 3 as it is one of the main parts of this thesis.

1.5.2   Pulse Shaping

The same filters used in the transmitter are used in the receiver for both GFSK and DPSK signals as it acts as a low-pass filter i.e. the same filter coefficients are used.

1.5.3   De-Mapping

Two basic methods for detectors are the correlation method and the matched filter method; these can then either be coherent or non-coherent. In this thesis the coherent correlation method is used for GFSK, DQPSK and 8DPSK.Coherent detection implies that the receiver exploits knowledge of the carrier's phase to detect the signal, and with non-coherent detection the receiver does not utilize phase information.

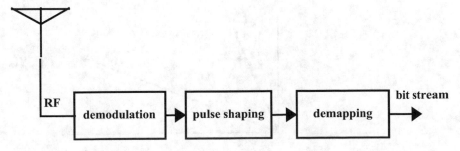

Fig. 1.7 Digital communication receiver

The decoding of the bit stream involves the symbol decision. For GFSK signals the symbol decision algorithm is determined according to the sign of the received symbol if it is positive so it was one else it was zero. For DPSK including both $\pi/4$ DQPSK and 8DPSK, the bits are decided by looking up in a table corresponding to the modulation scheme. These tables are the same as used in the modulation encoding.

1.6   CORDIC Theory

CORDIC (Coordinate Rotation Digital Computer) is a method for computing elementary functions using minimal hardware such as shifts adds/subs and compares. CORDIC works by rotating the coordinate system through constant angles until the angle is reduces to zero. The angle offsets are selected such that the operations on X and Y are only shifts and adds [39].

1.6.1   Introduction

The CORDIC algorithm performs a planar rotation. Graphically, planar rotation means transforming a vector (Xi, Yi) into a new vector (Xj, Yj) as shown below in Fig. 1.8.

Using a matrix form, a planar rotation for a vector of (Xi, Yi) is defined as

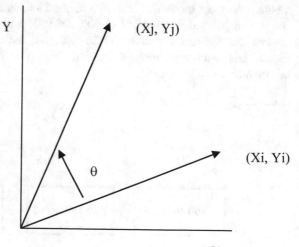

Fig. 1.8 CORDIC theory

$$\begin{bmatrix} X_j \\ Y_j \end{bmatrix} = \begin{bmatrix} \cos\theta & -\sin\theta \\ \sin\theta & \cos\theta \end{bmatrix} \begin{bmatrix} X_i \\ Y_i \end{bmatrix} \tag{1.11}$$

The θ angle rotationcan be executed in several steps, using an iterative process. Each step completes a small part of the rotation. Many steps will compose one planar rotation. A single step is defined by the following equation.

$$\begin{bmatrix} X_{n+1} \\ Y_{n+1} \end{bmatrix} = \begin{bmatrix} \cos\theta_n & -\sin\theta_n \\ \sin\theta_n & \cos\theta_n \end{bmatrix} \begin{bmatrix} X_n \\ Y_n \end{bmatrix} \tag{1.12}$$

Equation 1.11 can be modified as follow:

$$\begin{bmatrix} X_{n+1} \\ Y_{n+1} \end{bmatrix} = \cos\theta_n \begin{bmatrix} 1 & -\tan\theta_n \\ \tan\theta_n & 1 \end{bmatrix} \begin{bmatrix} X_n \\ Y_n \end{bmatrix} \tag{1.13}$$

Equation 1.12 requires three multiplies, compared to the four needed in Eq. 1.11. Additional multipliers can be eliminated by selecting the angle steps such that the tangent of a step is a power of 2. Multiplying or dividing by a power of 2 can be implemented using a simple shift operation. The angle for each step is given by:

$$\theta_n = \arctan\left(\frac{1}{2^n}\right) \tag{1.14}$$

All iteration-angles summed must equal the rotation angle θ.

$$\sum_{n=0}^{\infty} S_n\theta_n = \theta \tag{1.15}$$

This results in the following equation:

$$\tan\theta_n = S_n 2^{-n} \tag{1.16}$$

Combining Eqs. 1.12 and 1.15 results in:

$$\begin{bmatrix} X_{n+1} \\ Y_{n+1} \end{bmatrix} = \cos\theta_n \begin{bmatrix} 1 & -S_n 2^{-n} \\ S_n 2^{-n} & 1 \end{bmatrix} \begin{bmatrix} X_n \\ Y_n \end{bmatrix} \tag{1.17}$$

Rewrites $\cos\theta$ as:

$$\cos\theta_n = \cos\left(\arctan\left(\frac{1}{2^n}\right)\right) \tag{1.18}$$

Compute previous equation for all values of 'n' and multiplying the results, which we will refer to as K.

$$K = \frac{1}{P} = \prod_{n=0}^{\infty} \cos\left(\arctan\left(\frac{1}{2^n}\right)\right) \approx 0.607253 \qquad (1.19)$$

K is constant for all initial vectors and for all values of the rotation angle; it is normally referred to as the congregate constant. The derivative P (approx. 1.64676) is defined here because it is also commonly used [39, 40].

We can now formulate the exact calculation the CORDIC performs.

$$\begin{cases} X_j = K(X_i \cos\theta - Y_i \sin\theta) \\ Y_j = K(Y_i \cos\theta + X_i \sin\theta) \end{cases} \qquad (1.20)$$

Because the coefficient K is pre-computed and taken into account at a later stage, Eq. 1.16 may be written as:

$$\begin{bmatrix} X_{n+1} \\ Y_{n+1} \end{bmatrix} = \begin{bmatrix} 1 & -S_n 2^{-n} \\ S_n 2^{-n} & 1 \end{bmatrix} \begin{bmatrix} X_n \\ Y_n \end{bmatrix} \qquad (1.21)$$

Abbreviation of Eq. 1.21 yields to:

$$\begin{cases} X_{n+1} = X_n - S_n 2^{-2n} Y_n \\ Y_{n+1} = Y_n + S_n 2^{-2n} X_n \end{cases} \qquad (1.22)$$

At this point a new variable called 'Z' is introduced. Z represents the part of the angle ϴ which has not been rotated yet where:

$$Z_{n+1} = \theta - \sum_{i=0}^{n} \theta_i \qquad (1.23)$$

For every step of the rotation Sn is computed as a sign of Zn.

$$S_n = \begin{cases} -1 & if \ Z_n < 0 \\ +1 & if \ Z_n \geq 0 \end{cases} \qquad (1.24)$$

This algorithm is commonly referred to as driving Z to zero. The CORDIC core computes:

$$[X_j, Y_j, Z_j] = [P(X_i \cos(Z_i) - Y_i \sin(Z_i)), \ P(Y_i \cos(Z_i) + X_i \sin(Z_i)), \ 0] \qquad (1.25)$$

There's a special case for driving Z to zero:

$$X_i = \frac{1}{P} = K \approx 0.60725 \tag{1.26}$$

$$Y_i = 0 \tag{1.27}$$

$$Z_i = \theta \tag{1.28}$$

$$[X_j, Y_j, Z_j] = [\cos\theta, \quad \sin\theta, \quad 0] \tag{1.29}$$

Another scheme which is possible is driving Y to zero which is called vectoring mode CORDIC algorithm. This CORDIC core type computes:

$$[X_j, Y_j, Z_j] = \left[P\sqrt{X_i^2 + Y_i^2}, \quad 0, \quad Z_i + \arctan\left(\frac{Y_i}{X_i}\right) \right] \tag{1.30}$$

For this scheme:

$$X_i = X \tag{1.31}$$

$$Y_i = Y \tag{1.32}$$

$$Z_i = 0 \tag{1.33}$$

$$[X_j, Y_j, Z_j] = \left[P\sqrt{X_i^2 + Y_i^2}, \quad 0, \quad \arctan\left(\frac{Y_i}{X_i}\right) \right] \tag{1.34}$$

1.6.2 Summary of CORDIC Algorithm

Table 1.7 summarize Equations and results for rotation and vectoring modes of the CORDIC algorithm.

Table 1.7 Results for CORDIC algorithm

Rotation mode	Vectoring mode
Equations	
$X_{n+1} = X_n - S_n 2^{-2n} Y_n$ $Y_{n+1} = Y_n + S_n 2^{-2n} X_n$ $Z_{n+1} = Z_n - S_n \arctan(2^{-2n})$ $s_i = \begin{cases} -1 & \text{if } zi < 0 \\ +1 & \text{otherwise} \end{cases}$	$X_{n+1} = X_n - S_n 2^{-2n} Y_n$ $Y_{n+1} = Y_n + S_n 2^{-2n} X_n$ $Z_{n+1} = Z_n - S_n \arctan(2^{-2n})$ $s_i = \begin{cases} -1 & \text{if } yi < 0 \\ +1 & \text{otherwise} \end{cases}$
Results after n iterations	
$y_n = A_n(x_0 \cos z_0 - y_0 \sin z_0)$ $y_n = A_n(y_0 \cos z_0 + x_0 \sin z_0)$ $z_0 = 0$ $A_n = \prod_n \sqrt{1 + 2^{-2i}}$	$x_n = A_n \sqrt{x_0{}^2 + y_0{}^2}$ $y_0 = 0$ $z_n = z_0 + \arctan\left(\frac{y_0}{x_0}\right)$ $A_n = \prod_n \sqrt{1 + 2^{-2i}}$

Table 1.8 Comparison between LUT and CORDIC implementation

Using LUT	CORDIC
Larger memory elements for high resolution	Low gate count for high resolution
Low latency	High latency because of iterative method

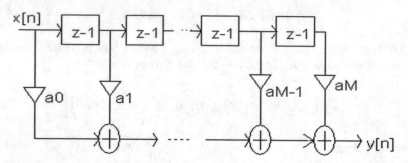

Fig. 1.9 FIR digital filter flow graph

Table 1.8 summarizes comparison between (look-up table) LUT and CORDIC implementation from which we can see that implementation of CORDIC algorithm is more area effective than LUT.

1.7 Digital Filters

Finite Impulse Response (FIR) filters possess many virtues, such as exact linear phase property, guaranteed stability, free of limit cycle oscillations, and low coefficient sensitivity However, the order of an FIR filter is generally higher than that of a corresponding infinite impulse response (IIR) filter meeting the same magnitude response specifications. Thus, FIR filters require considerably more arithmetic operations and hardware components such as delay, adder and multiplier [41].

This makes the implementation of FIR filters, especially in applications demanding narrow transition bands, very costly. When implemented in VLSI (Very Large Scale Integration) technology, the coefficient multiplier is the most complex and the slowest component. The cost of implementation of an FIR filter can be reduced by decreasing the complexity of the coefficients; coefficient complexity reduction includes reducing the coefficient word length [42].

An FIR digital filter of order M may be implemented by programming the signal flow graph shown below in Fig. 1.9 where its difference equation is:

$$Y[n] = a_0 x[n] + a_1 x[n-1] + a_2 x[n-2] + \ldots + a_m x[n-M] \qquad (1.35)$$

Where:

X (n) represents the sequence of input samples.

a_m represents the filter coefficients.

1.7.1   Advantages of Digital Filters

The following list gives some of the main advantages of digital over analog filter [43]. FIR digital Filter Frequency response is shown in Fig. 1.10.

(a) A digital filter is programmable; this means the digital filter can easily be changed without affecting the hardware. An analog filter can only be changed by redesigning the filter circuit.
(b) The characteristics of analog filter circuits (particularly those containing active components) are subject to drift and are dependent on temperature. Digital filters do not suffer from these problems, and so are extremely stable with respect both to time and temperature.
(c) Unlike their analog counterparts, digital filters can handle low frequency signals accurately.
(d) Digital filters are very much more versatile in their ability to process signals in a variety of ways; this includes the ability of some types of digital filter to adapt to changes in the characteristics of the signal.
(e) Digital filters can be easily designed and tested.

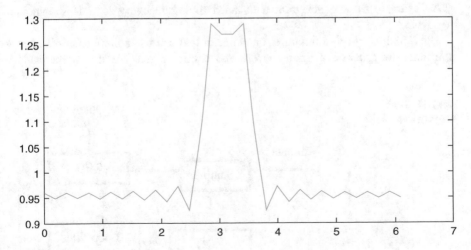

Fig. 1.10 FIR digital filter frequency response

1.8   Symbol Timing Recovery

Timing recovery is one of the most critical demodulator functions. Timing recovery algorithm can be classified into two categories according to the data dependency [44]:

- Class 1: Decision Directed (DD) or Data-Aided (DA).
- Class 2: Non-data-Aided (NDA).

In a receiver, the received signal is first demodulated and low-pass filtered to recover the message signals. The next step for the receiver is to sample the message signals at the symbol rate and decide which symbols were sent. Although the symbol rate is typically known to the receiver, the receiver does not know when to sample the signal for the best noise performance. The objective of the symbol-timing recovery loop is to find the best time to sample the received signal to obtain symbol synchronization. Two quantities must be determined by the receiver to achieve symbol synchronization. The first is the sampling frequency. Locking the sampling frequency requires estimating the symbol period so that samples can be taken at the correct rate. Although this quantity should be known, oscillator drift will introduce deviations from the stated symbol rate. The other quantity to determine is sampling phase. Locking the sampling phase involves determining the correct time within a symbol period to take a sample.

Real-world symbol pulse shapes have a peak in the center of the symbol period. Sampling the symbol at this peak results in the best signal-to-noise-ratio and will ideally eliminate interference from other symbols. This type of interference is known as inter-symbol interference.

The data aided timing recovery depends on a synchronization word sent by the transmitter in contrary to non-data aided timing recovery which does not need synchronization word and depends on feedback technique.

An example of a typical non data aided timing recovery loop is shown in Fig. 1.11.

The signal is sent to a timing error estimator that can use a number of different algorithms to generate a timing error. The control signal for the Numerically

Fig. 1.11 Timing recovery loop

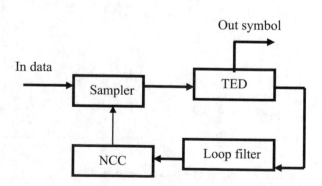

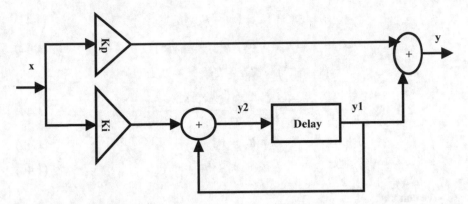

Fig. 1.12 Second order loop filter

Controlled Clock (NCC) is formed by filtering this error signal using a standard second order loop filter containing a proportional and an integral section [45].

An example of a typical second-order loop filter is shown in Fig. 1.12. The second-order loop filter consists of two paths. The proportional path multiplies the timing error signal by a proportional gain Kp.

From control theory, it is known that a proportional path can be used to track out a phase error; however, it cannot track out a frequency error. For a timing recovery loop to track out a sampling frequency error, a loop filter containing an integral path is needed. The integral path multiplies the error signal by an integral gain Ki and then integrates the scaled error using an adder and a delay block. A second- order filter can track out both a sampling phase and a sampling frequency error.

The timing error detector output, x, is multiplied by the proportional gain constant Kp in the upper arm. In the lower arm, the phase detector output is first multiplied by Ki, the integral gain constant. The result of this multiplication is fed into an integrator comprising an adder and a register (unit delay). The final output y is the sum of the product of the proportional gain constant K p computed in the upper arm, and the output of the integrator in the lower arm [46].

Here is a derivation of the transfer function y/x of this loop filter. I introduced a couple of intermediate variables, y_1 and y_2, to make the job easier:

$$y = K_p x + y_1 \tag{1.36}$$

$$y_1 = y - K_p x \tag{1.37}$$

$$y_1 = y_2 z^{-1} \tag{1.38}$$

$$y_2 = K_i x + y_1 \tag{1.39}$$

$$y_1 = (K_i x + y_1) z^{-1} \tag{1.40}$$

$$y_1 = (K_i x + y - K_p x) z^{-1} \tag{1.41}$$

$$y = K_p x + (K_i x + y - K_p x) z^{-1} \tag{1.42}$$

$$y(1 - z^{-1}) = K_p x + K_i x z^{-1} - K_p x z^{-1} \tag{1.43}$$

$$\frac{y}{x} = \frac{K_p + (K_i - K_p) z^{-1}}{(1 - z^{-1})} = \frac{b_0 + b_1 z^{-1}}{a_0 + a_1 z^{-1}} \tag{1.44}$$

$$K_p = b_0 \tag{1.45}$$

$$K_i = b_0 + b_1 \tag{1.46}$$

We can say:

$$y(n) = Kp * x(n) + Ki * x(n - 1) + y(n - 1) - Kp * x(n - 1) \tag{1.47}$$

Where $y(n)$ is the current filter output, $y(n-1)$ is the previous filter output, $x(n)$ is the current phase detector output, and $x(n-1)$ is the previous phase detector output.

The timing error detector output is computed and the filter output updated every Ts seconds, where Ts is the sampling interval. To compute b0 and b1, starting with an analog prototype and then using the bilinear z transform as shown in the following equations:

$$\tau_1 = \frac{K_o K_d}{w_n^2} \tag{1.48}$$

$$\tau_2 = \frac{2\xi}{w_n} \tag{1.49}$$

$$b_0 = \frac{T_s}{2\tau_1} \left(1 + \frac{1}{\tan\left(\frac{T_s}{2\tau_2}\right)} \right) \tag{1.50}$$

$$b_1 = -\frac{T_s}{2\tau_1} \left(1 - \frac{1}{\tan\left(\frac{T_s}{2\tau_2}\right)} \right) \tag{1.51}$$

In the equations above, K_o is the oscillator gain in radians sec^{-1} volt^{-1} in the analog domain, or radians sec^{-1} in the digital domain. K_d is the timing error detector gain in volts radians^{-1} in the analog domain, or unit radians^{-1} in the digital domain. The natural frequency of the loop is ω_n, in MHz and the damping factor is ζ. T_s is the sampling interval of the digital system, and τ_1 and τ_2 are the loop time constants where $\zeta = 0.7$, and K_o, K_d assumed to be equal 1.

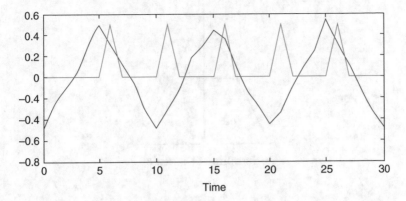

Fig. 1.13 Symbol sampling before DLL lock

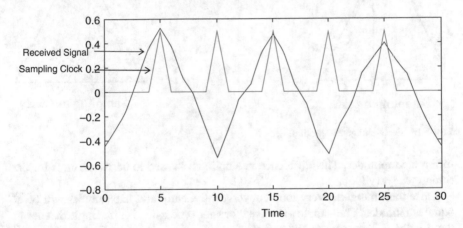

Fig. 1.14 Symbol sampling after DLL lock

The following figures can illustrate the timing recovery where Fig. 1.13 shows symbol sampling before delay locked loop (DLL) lock and Fig. 1.14 shows it after Locking.

1.8.1   Timing Error Detectors Algorithms

1.8.1.1   Early-Late Gate Algorithm

This timing recovery algorithm generates its error by using samples that are early and late compared to the ideal Sampling point. The generation of the error requires at least three samples per symbol.

The method of generating the error is illustrated in Fig. 1.15. The left plot is for the case where sampling is occurring late. Note that the early and late samples are at

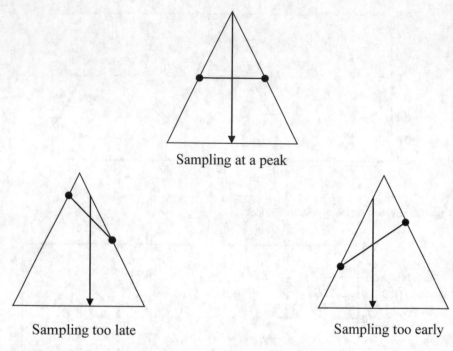

Sampling at a peak

Sampling too late   Sampling too early

Fig. 1.15 Early-late gate algorithm

different amplitudes. This difference in amplitude is used to derive an error for the timing recovery loop.

Once the timing recovery loop converges, the early and late samples will be at equal amplitudes. The sample to be used for later processing is the sample that lies in the middle of the early and late samples.

One drawback of the early-late gate algorithm is that it requires at least three samples per symbol. Thus, it is impractical for systems with high data rates.

1.8.1.2 Mueller and Muller Algorithm

The Mueller and Muller algorithm only requires one sample per symbol. The error term is computed using the following equation:

$$e_n = (y_n.y_{n-1}') - (y_n'.y_{n-1}) \tag{1.52}$$

Where y_n is the sample from the current symbol and y_{n-1} is the sample from the previous symbol. The slicer (decision device) outputs for the current and previous symbol are represented by y_n' and y_{n-1}', respectively.

Examples of the value for the Mueller and Muller error for the cases of different timing offsets are shown in Fig. 1.16. One drawback of this algorithm is that it is

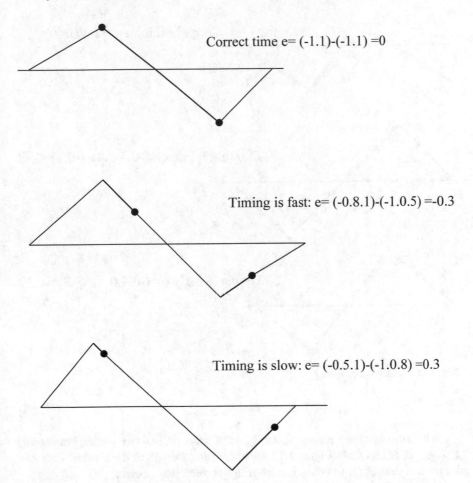

Correct time e= (-1.1)-(-1.1) =0

Timing is fast: e= (-0.8.1)-(-1.0.5) =-0.3

Timing is slow: e= (-0.5.1)-(-1.0.8) =0.3

Fig. 1.16 Mueller and Muller algorithm

sensitive to carrier offsets, and thus carrier recovery must be performed prior to the Mueller and Muller timing recovery.

1.8.1.3   Gardner Algorithm

The Gardner algorithm has seen widespread use in many practical timing recovery loop implementations. The algorithm uses two samples per symbol and has the advantage of being insensitive to carrier offsets. The timing recovery loop can lock first, therefore simplifying the task of carrier recovery.

The error for the Gardner algorithm is computed using the following equation:

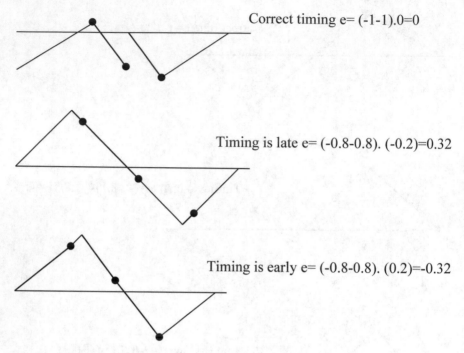

Correct timing e= (-1-1).0=0

Timing is late e= (-0.8-0.8). (-0.2)=0.32

Timing is early e= (-0.8-0.8). (0.2)=-0.32

Fig. 1.17 Gardner algorithms

$$e_n = (y_n - y_{n-2})y_{n-1} \qquad (1.53)$$

Where the spacing between y_n and y_{n-2} is T seconds, and the spacing between y_n and y_{n-1} is T/2 seconds. Figure 1.17 illustrates how the sign of the Gardner error can be used to determine whether the samplings correct, late or early.

Note that the Gardner error is most useful on symbol transitions (when the symbol goes from positive to negative or vice-versa). The Gardner error is relatively small when the current and previous symbol has the same polarity.

1.9   Carrier Recovery

The phase-locked loop (PLL) is a critical component in coherent communications receivers that is responsible for locking on to the carrier of a received modulated signal. Ideally, the transmitted carrier frequency is known exactly and we need only to know its phase to demodulate correctly. However, due to imperfections at the transmitter, the difference between the expected and actual carrier frequencies can be modeled as a time-varying phase. Provided that the frequency mismatch is small relative to the carrier frequency, the feedback control of an appropriately calibrated PLL can track this time-varying phase, thereby locking on to both the correct frequency and the correct phase. The goal of the PLL is to maintain a demodulating sine and cosine that match the incoming carrier.

There are two categories of techniques for carrier synchronization: assisted and blind. Assisted techniques multiplex a special signal called a pilot signal with the data signal. Blind synchronization techniques derive the carrier phase estimate directly from the modulated signal. This approach is more prevalent in practice and has the advantage that all of the transmitter power can be allocated to the information-bearing signal [47, 48].

1.9.1   Pilot Tone Assisted Carrier Recovery

A system that uses pilot tone assisted carrier recovery multiplexes an un-modulated tone with the information signal. The tone is commonly added at either the carrier frequency, as shown in Fig. 1.18, or a multiple of the carrier frequency. The receiver then uses a phase locked loop to acquire and track the carrier component. The phase locked loop is designed to have a narrow bandwidth so that the information signal does not significantly affect the recovered carrier estimate [24, 49–56].

1.9.2   Costas Loop

The Costas loop, developed in 1956, can recover the carrier from suppressed carrier signal. A block diagram is shown in Fig. 1.19. The received signal is multiplied by the recovered carrier and a version of the recovered carrier that has been shifted by 90 degrees. The results from each of these operations are low-pass filtered and multiplied together to generate an error signal that is used to drive a Voltage Controlled Oscillator (VCO).

Fig. 1.18 Signal with embedded Pilot Tone

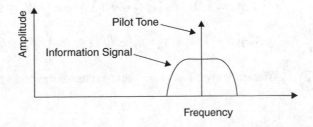

Receiced signal

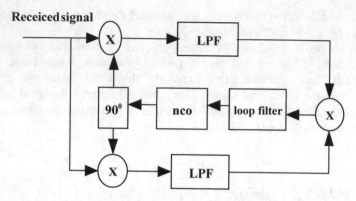

Fig. 1.19 Costas loop

1.9.3   Decision Feedback Phase Locked Loop

The Costas loop treat the information signal as a random sequence and use its statistical average to recover the carrier information embedded within it. A decision feedback PLL, assumes that the information sequence is known. The data sequence is obtained from the received signal by the Detector. Therefore, the phase contribution of each symbol can be removed in the phase detector before calculating an error signal once per symbol.

Suppose Wc is the believed digital carrier frequency. We can then represent the actual received carrier frequency as the expected carrier frequency with some offset.

$$\widetilde{W}_c = W_c + \widetilde{\theta}[n] \tag{1.54}$$

The NCO generates the demodulating sine and cosine with the expected digital frequency W_c and offsets this frequency with the output of the loop filter. The NCO frequency can then be modeled as:

$$\widetilde{W}_c = W_c + \theta[n] \tag{1.55}$$

Using the appropriate trigonometric identities, the in-phase and quadrature signals can be expressed as:

$$Z_I[n] = 1/2\left(\cos\left(\widetilde{\theta}[n] - \theta[n]\right) + \cos\left(2W_c + \widetilde{\theta}[n] + \theta[n]\right)\right) \tag{1.56}$$

$$Z_Q[n] = 1/2\left(\sin\left(\widetilde{\theta}[n] - \theta[n]\right) + \sin\left(2W_c + \widetilde{\theta}[n] + \theta[n]\right)\right) \tag{1.57}$$

After applying a low-pass filter to remove the double frequency terms, we have

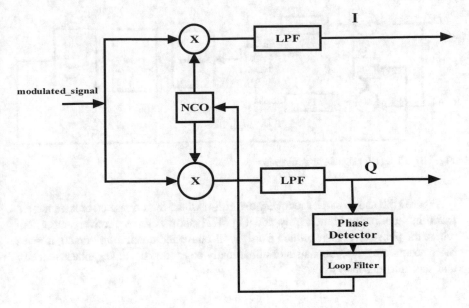

Fig. 1.20 PLL (Q-signal) block diagram

$$y_I[n] = 1/2 \cos \left(\widetilde{\theta}[n] - \theta[n] \right) \tag{1.58}$$

$$y_Q[n] = 1/2 \sin \left(\widetilde{\theta}[n] - \theta[n] \right) \tag{1.59}$$

Note that the quadrature signal is zero when the received carrier and internally generated waves are exactly matched in frequency and phase. When the phases are only slightly mismatched we can use the relation

$$\sin (\theta) \approx \theta \tag{1.60}$$

And let the current value of the quadrature channel approximate the phase difference:

$$y_Q[n] \approx \left(\widetilde{\theta}[n] - \theta[n] \right) \tag{1.61}$$

With the exception of the sign error, this difference is essentially how much we need to onset our NCO frequency. In a more advanced receiver, information from both the in-phase and quadrature branches is used to generate an estimate of the phase error.

If $\left(\widetilde{\theta}[n] - \theta[n] \right) > 0$ then $\theta[n]$ is too large and we want to decrease our NCO phase.

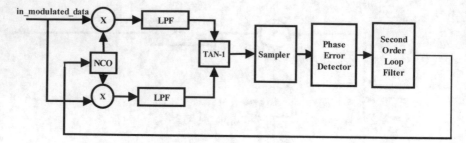

Fig. 1.21 PLL (1-Q) signals bock diagram

Figure 1.20 shows carrier recovery algorithm where their phase error is generated from the quadrature branch only and Fig. 1.21 shows carrier recovery algorithm where the phase error is generated from the in-phase and quadrature branch, it gives more accurate results, it consists of numerically controlled oscillator, phase detector and second order loop filter.

References

1. Specification of the Bluetooth System, Core Specifications, version 2.0+EDR, v1.2, v1.1; https://www.bluetooth.org/spec/
2. P. McDermott-Wells, What is Bluetooth? IEEE Potentials **23**(5), 33–35 (2005)
3. K.V.S.S.S.S. Sairam, N. Gunasekaran, S.R. Redd, Bluetooth in wireless communication. IEEE Commun. Mag. **40**(6), 90–96 (2002)
4. B. Chatschik, An overview of the Bluetooth wireless technology. IEEE Commun. Mag. **39**(12), 86–94 (2001)
5. M. Cominelli, P. Patras, F. Gringoli, One GPU to snoop them all: a full-band Bluetooth low energy sniffer. In *Proc. of IEEE MedComNet'20*, Jun. 2020, pp. 5–8
6. S. Han, Y. Park, H. Kim, Extending Bluetooth le protocol for mutual discovery in massive and dynamic encounters. IEEE Trans. Mobil Comput. **18**(10), 2344–2357 (2019)
7. Bluetooth SIG, Specification of the Bluetooth System, http://www.bluetooth.Com/developer/specification/Bluetooth_11_Specifications_Book (Apr. 9, 2001)
8. J. Martin, D. Alpuche, K. Bodeman, L. Brown, E. Fenske, L. Foppe, T. Mayberry, E. Rye, B. Sipes, S. Teplov, Hando_ all your privacy—a review of Apple's Bluetooth low energy continuity protocol. Proc. Priv. Enhancing Technol. **2019**, 34–53 (2019)
9. P.P. Ray, S. Agarwal, Bluetooth 5 and internet of Things: potential and architecture. In *Signal Processing, Communication, Power and Embedded System (SCOPES), 2016 International Conference on*. IEEE, 2016, pp. 1461–1465
10. S. Raza, P. Misra, Z. He, T. Voigt, Building the Internet of Things with Bluetooth smart. Ad Hoc Netw. **57**, 19–31 (2017)
11. P. Karthikeyan, N. Sumanth, S. Jude, Bluetooth based function control in a car. IOP Conf. Ser. Mater. Sci. Eng. **263**, 42065 (2017)
12. A.M. Lonzetta, P. Cope, J. Campbell, B.J. Mohd, T. Hayajneh, Security vulnerabilities in Bluetooth technology as used in IoT. J. Sens. Actuator Netw. **7**, 28 (2018)
13. K.S. Mohamed, The era of Internet of Things: towards a smart world, in *The Era of Internet of Things*, (Springer, Cham, 2019)

14. B. Badihi, F. Ghavimi, R. Jäntti, On the system-level performance evaluation of Bluetooth 5 in IoT: Open office case study. In *2019 16th International Symposium on Wireless Communication Systems (ISWCS)*, Oulu, Finland, 2019, pp. 485–489

15. M. Spörk, C. A. Boano, K. Orme, Performance and trade-offs of the new PHY modes of BLE 5. In *Proceedings of the ACM MobiHoc Workshop on Pervasive Systems in the IoT Era – PERSIST-IoT '19*, Catania, Italy, July 2019, pp. 7–12

16. Bluetooth Core Specification v5.1. https://www.bluetooth.com/bluetooth-resources/bluetooth-core-specification-v5-1-featureoverview/. Online; accessed 04-December-2019 (2019)

17. http://electronics.howstuffworks.com/Bluetooth.html

18. W. Narzt, L. Furtmüller, M. Rosenthaler, Is Bluetooth low energy an alternative to near field communication? J. Mob. Multimed. **12**, 76–90 (2016)

19. C. Gomez, J. Oller, J. Paradells, Overview and evaluation of Bluetooth low energy: an emerging low-power wireless technology. Sensors **12**, 11734–11753 (2012)

20. Z. Zuo, L. Liu, L. Zhang, Y. Fang, Indoor positioning based on Bluetooth low-energy beacons adopting graph optimization. Sensors (Switzerland) **18**, 3736 (2018)

21. K.S. Mohamed, *New Frontiers in Cryptography* (Springer, USA, 2020)

22. F. Zafari, I. Papapanagiotou, K. Christidis, Microlocation for Internet-of-Things-equipped smart buildings. IEEE Internet Things J. **3**, 96–112 (2016)

23. A. MacKey, P. Spachos, Performance evaluation of beacons for indoor localization in smart buildings. In *Proceedings of the 2017 IEEE Global Conference on Signal and Information Processing*, Montreal, QC, Canada, 14–16 November 2017, pp. 823–827

24. K.H. Chang, Bluetooth: a viable solution for IoT? [industry perspectives]. IEEE Wirel. Commun. **21**(6), 6–7 (2014)

25. S. Raza, P. Misra, Z. He, T. Voigt, Bluetooth smart: an enabling technology for the Internet of Things. In *Wireless and Mobile Computing, Networking and Communications (WiMob), 2015 IEEE 11th International Conference on*, Oct 2015, pp. 155–162

26. M. Andersson, Use case possibilities with Bluetooth low energy in IoT applications, White Paper, 2014

27. W. Roy et al., Bluetooth LE finds its niche. IEEE Pervasive Comput. **12**(4), 12–309 (2013)

28. J. Lin, W. Yu, N. Zhang, X. Yang, H. Zhang, W. Zhao, A survey on internet of things: architecture, enabling technologies, security and privacy, and applications. IEEE Internet Things J. **4**(5), 1125–1142 (2017)

29. N.B. Suryavanshi, K. Viswavardhan Reddy, V.R. Chandrika, Direction finding capability in Bluetooth 5.1 standard, in *Ubiquitous Communications and Network Computing*, ed. by N. Kumar, R. Venkatesha Prasad, (Springer International Publishing, Cham, 2019), pp. 53–65. ISBN: 978-3-030-20615-4

30. M. Cominelli, P. Patras, F. Gringoli, Dead on arrival: an empirical study of the Bluetooth 5.1 positioning system. In *Proceedings of the 13th International Workshop on Wireless Network Testbeds, Experimental Evaluation & Characterization. WiNTECH '19*. Los Cabos, Mexico: ACM, 2019, pp. 13–20. ISBN: 978-1-4503-6931-2. URL: http://doi.acm.org/10.1145/3349623.3355475

31. SIG, Bluetooth. Bluetooth Core Specification v5.0. https://www.mouser.it/pdfdocs/bluetooth-Core-v50.pdf. Online; accessed 29-January-2020 (2016)

32. M. Collotta, G. Pau, A novel energy management approach for smart homes using Bluetooth low energy. IEEE J. Select. Areas Commun. **33**(12), 2988–2996 (2015)

33. R. Jacob et al., Synchronous transmissions on Bluetooth 5 and IEEE 802.15.4: a replication study. In *Proc. of the 3rd CPS-IoTBench Workshop*, 2020

34. Bluetooth Low Energy Beacons. SWRA475A. Revised Oct. 2016. Texas Instruments. Jan. 2015

35. Bluetooth Core Specification. v5.0. Bluetooth SiG. Dec. 2016

36. S. Sampei, *Applications of Digital Wireless Technologies to Global Wireless Communications* (Prentice Hall Inc., Upper Saddle River, 1997)

37. J. Iversen, L.P. Nissen, *Digital Down-Converter and Sample Rate Converter* (Master thesis, Aalborg University, June 2003)
38. A. Madisetti, Y. Kwentus, A. Willson, A 100-MHz, 16-b, direct digital frequency synthesizer with a 100-dBc spurious-free dynamic range. IEEE J. Solid State Circuits **34**(8), 1034–1043 (1999)
39. R. Andraka, A survey of CORDIC algorithms for FPGA base computers. In *Proceedings of the 1998 ACMISIGDA 6th International Symposium on Field Programmable Gate Arrays*, pp. 191–2000 (1998)
40. T.S. Jeng, *Design of A CORDIC Calculator* (Bachelor of Engineering, University of New South Wales, November 2004)
41. K.A. Vinger, J. Torresen, Implementing evolution of FIR- filters efficiently in an FPGA. *Proceeding NASA/DoD Conference on Evolvable Hardware*, 9–11 July 2003, pp. 26–29
42. K. Wiatr, Implementation of multipliers in FPGA structures. *International Symposium on Quality Electronic Design*, 26–28 March 2001, pp. 415–420
43. D. Kocur, *Digital Filter Lessons* (Undergraduate Engineering Courses at The technical University of Korsice, Slovakia, 2002)
44. W. Han, *The Application of Multirate Techniques for Synchronization in Fully Digital Demodulation* (1995)
45. D.L. Jones, W. Appadwedula, M. Berry, M. Haun, D. Moussa, D. Sachs, Digital Receivers: Symbol-Timing Recovery for QPSK. The Connexions Project, Feb 2004
46. W. Han, *The Application of Multirate Techniques for Synchronization in Fully Digital Demodulation* (Master thesis, Asian Institute of Technology Bangkok, Thailand, December, 1995)
47. J. Che, *Carrier Recovery in Burst-Mode 16-Qam* (Master thesis, 2004)
48. DL. Jones, W. Appadwedula, M. Berry, M. Haun, D. Moussa, D. Sachs, Digital Receiver: Carrier Recovery, The Connexions Project, Feb 2004
49. D. Aspel, *Adaptive Multilevel Quadrature Amplitude Radi Implementation in Programmable Logic* (University of Saskatchewan, April 2004)
50. K.S. Mohammed, FPGA implementation of PPM I-UWB baseband transceiver. *Proceedings of the European Computing Conference*. Springer, Boston, MA, 2009
51. K. Salah, Design and FPGA implementation of non-data aided timing and carrier recovery techniques for EDR Bluetooth standard. *Signal Processing Algorithms, Architectures, Arrangements, and Applications (SPA), 2008*. IEEE, 2008
52. K. Salah, FPGA implementation of Bluetooth 2.0 transceiver. *Proceedings of the 5th WSEAS International Conference on System Science and Simulation in Engineering*. World Scientific and Engineering Academy and Society (WSEAS), 2006
53. H.B. Pandya, T.A. Champaneria, Internet of Things: survey and case studies. In *2015 International Conference on Electrical, Electronics, Signals, Communication and Optimization (EESCO)*, Jan 2015, pp. 1–6
54. ABI Research, *Bluetooth 5 Evolution Will Lead to Widespread Deployments on the IoT Landscape* (USA, 2016)
55. T.S. Rappaport, *Wireless Communications: Principles and Practice* (Prentice Hall, USA, 2002)
56. Bluetooth Special Interest Group. 2016. Bluetooth Core Specifications. (Dec 2016). https://www.bluetooth.com/specifications/bluetooth-core-specification

Chapter 2
An Introduction to IoT

2.1 Introduction

Internet of Things (IoT) is a key technology driving toward a new perspective of the world where almost all conceivable things are connected to a network for remote sensing and control. The realization of the IoT is enabled by the integration of various technologies, including smart devices, wireless networking, cloud computing, and data analytics [1].

There is no single definition for Internet of Things (IoT). IoT is a new dimension of the internet and a new generation of services. IoT means anything can communicate with anything in any place at any time using any protocols. Any place because information are sent to the internet, so you can be access them from any place [2]. IoT is like a human society, with minimum human intervention as things have virtual identities to be known. IoT will enable "smart X", "X" can be anything such as TV, watch, glass, clock, coffee machine, and car. History of internet of things "IoT" back to 1997, but the first conference was launched on 2008. IoT AKA machine to machine "M2M", device to device "D2D", and "Ubiquitous".

In IoT, we start with a "Thing" and add computational intelligence to improve its function then add a network connection to further enhance its function [3]. Microsoft Azure is an example for a cloud computing platform. Cloud computing means that devices exchange information through a cloud infrastructure. Things in IoT can be physical or virtual.

By enabling IoT, control of daily life in an intelligent and easier way is feasible. There are many reasons that make IoT feasible such as: Internet infrastructure already exists and internet available almost everywhere in the developed world. Moreover, Hardware size allows incorporation into a device. Besides, cost of hardware has decreased. IPv6 protocol has large address space, so we can assign an IP for each thing on the earth. There is infinite number of applications for IoT such as: traceability, smart home, smart office, smart campus, smart shopping, and smart

K. S. Mohamed, *Bluetooth 5.0 Modem Design for IoT Devices*,
https://doi.org/10.1007/978-3-030-88626-4_2

clothes. IoT will not only control our homes, but also our business, society, cities, and our lives.

IoT is about intelligent not just control. IoT enabling technologies basically consists of four main functions: sensing, communication, control, and actuators which have a great analogy with human. A large number of industrial data, usually referred to big data [4], are collected from IoT. The IoT concept has been designed to perform several main distinctive actions: collect data, transfer/transmit/exchange data, change/process data, store data, and personalize/execute data.

Sensors can be real sensors or virtual sensors to collect data from the internet. Communication can be done using many types of protocols such as: RFID, AD-HOC, Ethernet, WIFI, 3G, 4G, Bluetooth, Zigbee, USB, WSN, and IPv6 which are ranging from short range to long range communications. For Example, Bluetooth for short range connectivity; WIFI for medium scale connectivity; cellular technologies for large scale connectivity. Control is done using FPGA, ASIC, or processors. Actuators examples are motor, alarm, and oven. IoT architecture consists of three layers: physical layer, communication layer, and application layer.

Nodes connected to each other using LANs which may or may not be connected to the internet (WAN) through gateways (using proxy to connect to the internet or without connectivity to provide intra connectivity between different LANs). Devices/Nodes are often connected to a Gateway in cases when the Device is not capable of directly connecting to further systems, e.g., if the Device cannot communicate via a particular protocol or because of other technical limitations. To solve these problems, a Gateway is used to compensate such limitations by providing required technologies and functionalities to translate between different protocols and by forwarding communication between Devices and other systems. A Gateway is, therefore, responsible for supporting the required communication technologies and protocols in both directions and for translating data if necessary. For instance, a Device communicates with a Gateway via an IoT protocol, such as ZigBee or MQTT. When the Gateway receives a message in a proprietary binary format from the Device, the Gateway translates the information into **JSON** or **XML** and forwards the data to a system in the World Wide Web. Likewise, the Gateway may translate commands into communication technologies, protocols, and formats supported by the respective Device. The Gateway may already execute some data processing functions, such as data aggregation, depending on its processing capabilities.

We can make full use of IoT technology when we overcome all its challenges and limitations. Any IoT system should satisfy 4 s's rule: simple, secure, smart, and scalable. Security and privacy are a challenge in IoT. Large number of IoT devices means increased threats, so a new security level is needed. We need to protect the cloud, the communication, and ensure privacy and integrity. In some applications, we need real time processing and maybe novel simulation techniques. Sensor reliability is an important limitation. There is no unified protocols and standardization for IoT. We need regulations to avoid multiple identities. Scalability is a challenge in IoT when we have massive number of devices. Low power and power harvesting is very important in IoT as most devices are battery-based devices [5].

Internet of Things (IoT) is a growing industry. Analysts predict that (IoT) products and services will grow exponentially in next years. It is a confluence of different sectors: embedded systems, communication systems, sensors/actuators, WWW and mobile applications. Use Internet of Things Technology to solve all Problems in different life sectors: health care, museums, libraries, inventory management, advertisement, real-estate identification, food tracking, maintenance, radiation/pollution monitoring, and security.

However, the Internet of Things is still maturing, in particular due to a number of factors, which limit the full exploitation of the IoT:

- No clear approach for the utilization of unique identifiers and numbering spaces for various kinds of persistent and volatile objects at a global scale.
- No standard IoT reference architecture.
- Less rapid advance in semantic interoperability for exchanging sensor information in heterogeneous environments.
- Difficulties in developing a clear approach for enabling innovation, trust and ownership of data in the IoT while at the same time respecting security and privacy in a complex environment. Difficulties in developing business which embraces the full potential of the Internet of Things. In our proposal, we believe that presenting simple yet clear services which can be mixed together to create complex scenarios to interact and react with Things around us.
- Government regulations on usage of GPS in (Geolocation) and restrictions on communication systems that interfere with police and military sector. In our proposal, we can easily switch between WIFI and 3G/GPRS. We believe that these regulations will be relaxed in future as situations get better and technology evolution imposes the change.

2.2   IoT Physical Layer

IoT is an extension of network connectivity to physical devices, such as actuators, sensors and mobile devices, enabled to interact and communicate among themselves, and can be controlled or monitored remotely. The physical layer is the most detailed level of abstraction in IoT. It mainly consists of sensors that acquire information for the system and actuators that do actions in response to instructions from the system. To imagine how they both, actuators and sensors, act together in a system, a smart house is considered for example. The actuators here are used to lock and unlock doors, switch on/off the lights and alert users of any warnings or control the temperature of a room or the whole house. The sensors are used to send feedback to the controller of each small system of those systems mentioned above. For example, they send feedback about the condition of the rooms and whether there are any people in the rooms or not, and accordingly, the controller sends its signals to the actuators to turn off unnecessary working devices such as the lights, the air conditioner... etc. Transducer terminology is used for both sensors and actuators. It means a device that converts energy form to another. In this chapter, different types of sensors and actuators are thoughtfully presented and discussed. Actuators may be written, sensors may be read. Moreover, different controllers used in IoT are discussed with its programming methods [6–8].

One of the most essential components for IoT is the sensors. Sensors are basically sense the physical phenomena or property that happens around them and sense different parameters according to the purpose of usage such as temperature, pressure, and humidity. Each sensor can only measure a unique property. Sensors are manufactured in different shapes and sizes. They can be mechanical sensors, electrical sensors, and chemical sensors. Sensors do not affect the measured property. The sensors can be classified according to their output as analog or digital sensors or according to the data type (scalar or vector). Moreover, they are divided in to Active and Passive. Active sensors are those which require an external excitation signal or a power signal. Passive Sensors, on the other hand, do not require any external power signal and directly generates output response.

One of the most **sensors requirements** are accuracy, resolution, and sensitivity. Most sensors have linear behavior. A sensor is a tiny device that measures a specific physical quantity. All IoT systems depend on the existence of one or more sensors. They are very essential in all aspects of life; as they are considered a feedback to the control that gives its signal to the actuator to reach a desired goal. There are different types of sensors, including phone-based, medical, environmental, and chemical sensors. They all have light weights and single functions, in addition to being inexpensive and miniaturized devices, but constrained to the battery capacity and the ease of deployment [9].

One of the most essential components for IoT is the actuators. Actuators are basically performing some actions based on the readings of the sensors and the required specifications which differ from an application to another. An actuator requires a control signal and a source of energy. There are three different types of actuators: mechanical, electrical, and pressure. Actuators convert energy to motion. An actuator is a device that converts an electrical signal into a mechanical signal or any other useful form of energy. Some examples include speakers, heaters, cooling elements and displays. They can be electrical, hydraulic or pneumatic actuators depending on their theory of operation. For example, hydraulic actuators use fluid mechanics to facilitate motion, whereas pneumatic actuators make use of the compressed air to generate pressure difference [10].

The controller is the device that receives the sensors' signals, processes them and makes computations on them, and then sends instruction signals to the actuators. Usually in control systems, these instruction signals are based on the difference between the sensors' readings and the desired values of the physical quantities, and thus these instruction signals are sent to the actuators in order to set the system back to the desired physical quantities' values. There are many hardware platforms with different capabilities that can be used in IoT applications. Choosing which hardware platform is used is based on the requirement of the IoT applications and depends also on whether we need it for development only or mass production [7, 8, 11].

In order for the hardware to perform well, operating systems should be installed. Operating systems organize the usage of hardware. For IoT applications, low power and small hardware overhead operating systems should be used.

The software platform is necessary to recognize the received data, identify the needed manipulation for the desired action by the user and transmit efficiently the new data to the right node.

There are multiple of business operating systems such as IBM Watson platform as well as open source platforms such as Linux, RIOT …etc. Choosing the right operating system is a crucial move in order to build the optimal IoT system for the desired application. In this section, the key parameters to choose the suitable operating system (OS) are investigated as follows:

- IoT heterogeneous hardware support: A lot of IoT systems usually work on different types of hardware from 16-bit microcontrollers to FPGAs based on the implemented hardware. Therefore, the operating system shall be compatible with the implanted hardware platform in order to achieve excellent performance.
- Real-time operating systems: One of the most important factors that guide most of IoT designs is that whether the operating system supports predictability or not. Predictability allows the system to be in an alert position based on earlier received data. This helps the software take actions rapidly, especially in situations like fires and accidents on the road. In turn, this is great evidence on how predictability reflects the degree of smartness of IoT systems, which is in high need for development.
- Developer friendly: IoT evolves rapidly, which requires an easy platform to apply new solutions and add smart features to the desired applications. In order to have low time-to-market, it is required to decide what operating systems are sufficient for developing the desired IoT system. Working on a developer friendly platform is highly required for companies in order not to be delayed for the market, which has its drawbacks.
- Memory: The operating system shall have low size in order to fit in the internal memories of microcontrollers. On the other hand, extra memory is needed to fulfill the size of the operating system.
- Security: Security is a very important factor that should be considered in all the layers of IoT. In order to achieve a good level of security, the operating system should support encryption engines, secure boot functions and usage of wireless authentication protocols.
- Accommodation for low power: Despite the fact that the hardware platform takes power issues into account, operating systems that allow power management capabilities gain much focus from IoT designers. Power management feature helps increase the battery life, especially for end-nodes. For example, the software can operate only with the required area of the hardware for simple functions as long as there is no necessity to perform repeated calculations, which is the case in most times. Focusing on such a thing increases the reliability of IoT products.

- Support required for communication and network: IoT requires different types of protocols in small, medium and large signals. The software has to strongly support such integration.

2.3   IoT Network and Communication Layer

The communication layer is considered as the backbone of the IoT systems. It is the main channel between the application layer and different operating activities in the IoT system. The whole physical system is loaded with amounts of data and information that need to be shared with other nodes. Therefore, it is needed to set up a suitable connection network among these nodes through a communication protocol. The communication could be wire-connected or wireless based on the protocol defined by the designer. Moreover, networks are very vital components in IoT to connect things to the outside world of internet. IoT requires an intelligent network infrastructure. The requirements of IoT communications includes: energy efficiency, range, cost, reliability, security, delay and scalability. The Open Systems Interconnection (OSI) model is an ISO-standard abstract model that describes a stack of seven protocol layers. From the top down, these layers are: application, presentation, session, transport, network, data link and physical. TCP/IP model includes only four layers, merging some of the many emerging and competing networking technologies are being adopted within the IoT space [27]. Multiple technologies are offered by different vendors or are aimed at different vertical markets like home automation, healthcare, or industrial IoT; often provide alternative implementations of the same standard protocols [12, 13].

Bluetooth is a wireless communication protocol widely used nowadays to connect devices together. It was first introduced in 1989 by Nick Rydbeck and Johan Ullman. It uses electromagnetic (EM) waves with frequencies ranging mainly between 2.4 GHz and 2.485 GHz. It is based on a master-slave configuration in which communication is established between a master and up to seven slaves maximum. The latest version is Bluetooth 5, which supports higher Bluetooth speed connections (2 Mbits/s) and further range (more than 300 m). Bluetooth 5 also contains features that support IoT, such as coded communication and forwarded error correction.

There are many advantages for Bluetooth that makes it suitable for IoT use. First of all, it is a wireless protocol, and thus it supports wireless applications such as wearable devices. Secondly, Bluetooth is a low power protocol, which is also ideal for IoT. On the other hand, Bluetooth limits the number of slave devices connected to a master device (7 devices maximum). In addition, power transmission limits the maximum distance at which a connection could be established, which creates a tradeoff between power consumption and furthest distance for communication [14–21]. Figure 2.1 shows Bluetooth devices example.

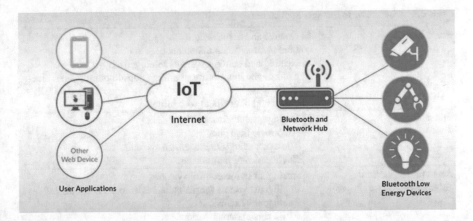

Fig. 2.1 Bluetooth devices

2.4 IoT Application Layer

Internet of things' application layer holds the responsibility for providing services and defines the set of messages' protocols that are passing at this level. There must be some data processing environment for analyzing the data fetched from the devices (sensors, controllers. . . . etc.) and making this data usable. Thus, this usability can be through direct applications with easy graphical user interface (GUI) for individual users, like mobile applications for simple IoT applications, or for massive projects that host global users, clouds can be used to analyze, sort out and store the data, and websites can be used as an interface. The application layer is also concerned with providing a virtual service layer that is responsible for data transport, security, and service discovery and device management on a high level of abstraction, independent of communication technologies in the lower layers. This ensures the right connectivity between devices and various IoT applications to realize horizontally integrated IoT for specific applications. This virtual service layer provides information collected from objects and the performance of the actuators. For instance, while data from a temperature sensor for home automation are provided, it should also describe if it is the indoor temperature of a room, or a fridge. . . etc. IoT potential allows it to customize any kind of applications. Table 2.1 summarizes some of the IoT applications [22–29].

2.5 AI, Big Data and IoT

By 2022, data will move from being Non-human centric to be human-centric data. Huge data are created everyday on social media, these huge data are called big data. Moreover, things that are connected to the internet will exceed number of population on each as it is expected to be 50 Billion devices by 2022. Human to Thing

Table 2.1 Domain of IoT usage: applications

Smart homes	Control and home security. Intelligent systems maintenance. Heating and cooling systems intelligent. Control and monitoring of energy consumption (water, electricity, gas). Facial and biomedical recognition.
Smart cities	Intelligent monitoring. Automatic transport. The exact energy management systems. Environmental monitoring.
Smart transportation/ automotive	Intelligent traffic control systems. Intelligent systems for maintenance of roads (land, air and sea). Intelligent systems parking. RFID tags communication.
Smart retail and logistics	Supply chain control. Intelligent shopping applications. Smart product management. Inventory tracking. Point-of-sale terminals. Vending machines.
Smart agriculture	Sensors check the soil moisture and temperature: Soil moisture management. Smart irrigation. Smart dust.
Smart factories and industries/ business	Indoor air quality. Temperature monitoring. Ozone presence. Indoor location. Vehicle auto-diagnosis. Sensors check the soil moisture and temperature.
Smart health care	Patients surveillance. Sportsmen care. Ultraviolet radiation. Smart hospitals.
Smart wearable	Smart glasses. Smart clothes. Sleep sensor. Smart watch.
Others	Smart museums. Smart schools. ATMs.

interaction will be anywhere, at any anytime. Computing devices are embedded in things or objects. Low-power and low-data rate wireless communication networks are the basic infrastructure for IoT. The **latency** of any IoT system (propagation delay +channel delay+ processing delay) can be small by implementing Edge/Fog computing principles. When considering a human analogy, radio communications in the cyber-physical fusion (AI, Big Data, and IoT) corresponds to the role of the nervous system that transmits information between the brain, i.e., AI, and each organ, i.e., device, such as the eyes and limbs [30].

Cloud computing means storing and accessing data and programs over the Internet instead of your computer's hard drive. The cloud is just a metaphor for the Internet. Cloud computing is a shared pool of computing/storage resources that can be accessed on demand and dynamically offered to the user. Cloud computing services can be accessed at any time from any place. They are offered by many companies such as Google (AWS) and Microsoft (Azure). Moreover, there are open platforms for cloud computing such as Thingspeak and Thingsboard. IoT includes heavy data (**big data**) transactions that need to be stored and analyzed.

Cloud service offerings are divided into three categories: infrastructure as a service (**IaaS**), platform as a service (**PaaS**), and software as a service (**SaaS**). IaaS is responsible for managing the hardware, network and other services. PaaS supports the OS and application platform, and SaaS supports everything. They are summarized in Table 2.1. A cloud system may be public or private. Public clouds can be accessed by anyone. Private clouds offer services to a set of authorized users. A hybrid cloud is a combination of both public and private clouds. In the IoT, sensors and devices communicate with one another and send data to the cloud for storage.

A cloud server is an open logical server which builds, hosts, and delivers through a cloud-computing platform through the Internet. Cloud servers maintain and exhibit similar functionality as well as capabilities to a typical server. However, they are accessing remotely from a cloud service provider as open server. A cloud server known as a virtual private server or virtual server. A cloud server is an Infrastructure as a Service (IaaS) primarily based cloud service model.

There are mainly two types of **cloud servers: logical and physical**. A cloud server is a logical when it delivered through server virtualization. The physical server is distributing into two or more logical servers, each of which has a separate OS, user interface and apps, although they share physical components from the underlying physical server, in this delivery model. At the same time, the physical cloud server is too accessing through the Internet remotely, it is not shared nor distributed. And it is commonly known as a dedicated cloud server.

Cloud computing is an example of an information technology(IT) paradigm, a model for enabling ubiquitous access to shared pools of configurable resources (such as computer servers, storage, applications and services, and networks), which can be rapidly provisioned with minimal management efforts, over the internet. Cloud computing is basically allows enterprises with various computing capabilities to store and it process data either in a privately-owned cloud or on a third-party server located in a data center thus making data-accessing mechanisms more efficient as well as reliable. To achieve coherence and economy of scale, similar to a utility cloud computing relies on sharing of resources.

Cloud computing allows other companies to minimize or avoid up-front IT infrastructure costs. On another hand, instead of wasting resources on computer infrastructure and their maintenance third-party clouds enable organizations to focus on their core businesses. Cloud providers coherently use a pay as you model. This might lead to unexpectedly high charges only if administrators are not familiarizing with cloud-pricing models [31].

Data from various sources and domains produced by IoT is often data stream, such as numeric data from different sensors or social media text inputs. Common data streams generally follow the Gaussian distribution over a long period. However, IoT data are produced in short time and in large quantities, presenting a variety of sporadic distributions over time. In addition, in some cases, in real time or near real time. One trend in Internet applications of Things that addresses the concept of IoT Analytics is the use of Fog Computing that can decentralize the processing of IoT data streams and only perform the transfer of filtered IoT data from the devices from the edge of the network to the cloud. Therefore, the association of data stream analysis with Fog Computing allows companies to explore in real time the data produced by IoT and thus produce business value [32].

2.6 Conclusions

This chapter presents a comprehensive survey of IoT. Moreover, it presents the different fundamental and basic concepts for IoT and its overall operation. This chapter discusses the IoT different layers: physical layer, network layer, and application layer. Also, it discusses big data and its relation to IoT.

References

1. K.E. Jeon, J. She, P. Soonsawad, P.C. Ng, BLE beacons for Internet of Things applications: survey, challenges, and opportunities. IEEE Internet Things J. **5**(2), 811–828 (2018)
2. I. Yaqoob, E. Ahmed, I.A.T. Hashem, A.I.A. Ahmed, A. Gani, M. Imran, M. Guizani, Internet of Things architecture: recent advances, taxonomy, requirements, and open challenges. IEEE Wirel. Commun. **24**(3), 10–16 (2017)
3. S.K. Tayyaba, M.A. Shah, O.A. Khan, A.W. Ahmed, Software Defined Network (SDN) based Internet of Things (IoT): a road ahead. In *Proceedings of the International Conference on Future Networks and Distributed Systems*. ACM, July 2017, pp. 1–8
4. M. Mohammadi, A. Al-Fuqaha, S. Sorour, M. Guizani, Deep learning for IoT big data and streaming analytics: a survey. IEEE Commun. Surv. Tutor. **20**(4), 2923–2960 (2018)
5. A. Dorri, S.S. Kanhere, R. Jurdak, P. Gauravaram, Blockchain for IoT security and privacy: the case study of a smart home. In *Proceedings of the IEEE International Conference on Pervasive Computing and Communications (PerCom) Workshops*. IEEE, March 2017, pp. 618–623
6. A. Al-Fuquay et al., Internet of Things: a survey on enabling technologies protocols and applications. IEEE Commun. Surveys Tut. **17**(4), 2347–2376 (2015)
7. "Products I IoT Solutions – ARM", ARM I The Architecture for the Digital World, 2016. [Online]. Available: https://www.arm.com/products/iot-solutions (Sept. 24 2016)
8. "IoT Hardware Guidebook I 2016 Prototyping Boards and Development Kits", Postscapes.com, 2016. [Online]. Available: http://www.postscapes.com/internet-of-things-hardware/ (Sept. 24 2016)

9. B. Ramachandran, IoE/IoT | Anything Connected, Connectedtechnbiz.wordpress.com, 2017. [Online]. Available: https://connectedtechnbiz.wordpress.com/category/ioeiot/. Accessed: 04 Jul 2017

10. K. Nair, J. Kulkarni, M. Warde, Z. Dave, V. Rawalgaonkar, G. Gore, J. Joshi, Optimizing power consumption in IoT based wireless sensor networks using Bluetooth low energy. In *Proc. Int. Conf. Green Comput. Internet Things (ICGCIoT)*, Oct. 2015, pp. 589–593

11. A. Al-Fuqaha et al., Internet of Things: a survey on enabling technologies protocols and applications. IEEE Commun. Surv. Tut. **17**(4), 2347–2376 (2015)

12. S.R. Pokhrel, C. Williamson, Modeling compound TCP over WiFi for IoT. IEEE/ACM Trans. Netw. **26**, 864–878 (2018)

13. https://developer.ibm.com/articles/iot-lp101-connectivity-network-protocols/

14. K.S. Mohammed, FPGA implementation of PPM I-UWB baseband transceiver, in *Proceedings of the European Computing Conference*, (Springer, Boston, 2009)

15. K. Salah, Design and FPGA implementation of non-data aided timing and carrier recovery techniques for EDR Bluetooth standard. *Signal Processing Algorithms, Architectures, Arrangements, and Applications (SPA), 2008*. IEEE, 2008

16. K. Salah, FPGA implementation of Bluetooth 2.0 transceiver. *Proceedings of the 5th WSEAS International Conference on System Science and Simulation in Engineering*. World Scientific and Engineering Academy and Society (WSEAS), 2006

17. K.H. Chang, Bluetooth: a viable solution for IoT? [industry perspectives]. IEEE Wirel. Commun. **21**(6), 6–7 (2014)

18. H.B. Pandya, T.A. Champaneria, Internet of Things: survey and case studies. In *2015 International Conference on Electrical, Electronics, Signals, Communication and Optimization (EESCO, USA)*, Jan 2015, pp. 1–6

19. ABI Research, *Bluetooth 5 Evolution Will Lead to Widespread Deployments on the IoT Landscape* (USA, 2016)

20. T.S. Rappaport, *Wireless Communications: Principles and Practice* (Prentice Hall, 2002)

21. Bluetooth Special Interest Group. 2016. Bluetooth Core Specifications. (Dec 2016). https://www.bluetooth.com/specifications/bluetooth-core-specification

22. Y. Mehta, J. Mounika, 5 Internet of Things (IoT) trends in 2017–19 – IoT worm, IoT worm, 2017. [Online]. Available: http://iotworm.com/5-internet-things-iot-trends-2017-19/. Accessed: 13 Aug 2017

23. Smart Traffic Management with Real Time Data Analysis, Cisco, 2017. [Online]. Available:

24. http://www.cisco.com/c/en_in/about/knowledge-network/smart-traffic.html. Accessed: 18 Jul 2017

25. J. Mounika, A. Prince, Y. Mehta, H. Nivas, A. Mahendra, Machine to Machine Communication Examples and Applications, IoT Worm, 2017. [Online]. Available: http://iotworm.com/machine-machine-communication-examples-applications/. Accessed: 13 Aug 2017

26. C. Key, 7 Best Developer Tools to Build your Next Internet of Things Application, Losant, 2017. [Online]. Available: https://www.losant.com/blog/7-best-developer-tools-to-build-your-next-internet-of-things-application. Accessed: 11 Jul 2017

27. L. Inc., "Losant", Losant, 2017. [Online]. Available: https://www.losant.com. Accessed: 11 Jul 2017

28. "IFTTT", Ifttt.com, 2017. [Online]. Available: https://ifttt.com. Accessed: 11 Jul 2017

29. Wireless: Programmable Cellular Data, SMS, and Voice for Devices – Twilio, Twilio.com, 2017. [Online]. Available: https://www.twilio.com/wireless. Accessed: 04 Jul 2017

30. R. Hadidi, J. Cao, M.S. Ryoo, H. Kim, Towards collaborative inferencing of deep neural networks on Internet of Things devices. IEEE Internet Things J. **7**(6), 4950–4960 (2020)

31. P. Lea, *Internet of Things for Architects: Architecting IoT Solutions by Implementing Sensors, Communication Infrastructure, Edge Computing, Analytics, and Security* (Packt, Birmingham, 2020)

32. L. Andrade, SOFT-IoT Platform in Fog of Things. WebMedia '18, October 16–19, 2018, Salvador-BA, Brazil

Chapter 3
Hardware Realization of GFSK-Based Bluetooth Modem

3.1 GFSK Transceiver Overview

Detection of GFSK can be based on relative frequency changes between symbol states and thus does not require absolute frequency accuracy in the channel. GFSK is thus relatively tolerant to local oscillator (LO) drift and Doppler Shift [1].

Everything in a GFSK modulator is the same as in a FSK modulator, except that before the baseband pulses go into the FSK modulator, it is passed through a Gaussian filter to make the pulse smoother and limits its spectral width. The spectral width for FSK is unlimited, in contrast; there is a limitation on GFSK [12–15].

One of the parameters that affect the spectrum is The $3 - dB$ bandwidth B of the pre-modulation filter is one of the parameters that control the bandwidth of the GFSK spectrum, decreasing the Bandwidth Symbol Period (BT) product where (T is the symbol period), results in a narrower GFSK spectrum, at the expense of a higher BER for the same Eb/N0 .This is caused by the increased level of inter-symbol-interference (ISI). Another parameter that affects the spectrum with is the modulation index h. Varying the modulation index effectively changes the signal separation at the sampling instants, this provides tradeoffs between bandwidth and BER [2, 16–18].

3.2 Survey of GFSK Transceiver Architecture

3.2.1 GFSK Transmitter Architecture

Since GFSK is basically a kind of frequency modulation various Frequency modulation techniques can be used to modulate the GFSK signal. Various techniques used to modulate the GFSK signal are briefly discussed in the next paragraphs [3–5].

K. S. Mohamed, *Bluetooth 5.0 Modem Design for IoT Devices*,
https://doi.org/10.1007/978-3-030-88626-4_3

3.2.1.1   Direct Modulation

In direct modulation Data is Gaussian filtered and fed to a Numerically Controlled Oscillator (NCO) which generates the frequency modulated signal. This GFSK NCO modulator architecture as shown in Fig. 3.1. In this arrangement the required modulation index of 0.35 is generated by the NCO. Unfortunately this type of modulation suffers from phase discontinuity which occurs at symbol boundaries and results in undesirable spectral characteristics as shown in Fig. 3.2.

3.2.1.2   Continuous Phase GFSK Modulation

Another technique is continuous phase GFSK modulation. The architecture of this modulation is shown in Fig. 3.3. This type of modulation doesn't suffer from the effects of phase discontinuities in contrast to direct modulation as shown in Fig. 3.4.

3.2.1.3   Quadrature Modulation

In quadrature modulation the data is Gaussian filtered and then integrated. This integrated data is then used to look up in the Sine/Cosine Look up table, which stores the modulated data as GFSK requires the generation of 2 symbols, one at a frequency $(\omega_c + \omega_1)$ and one at a frequency $(\omega_c - \omega_1)$. The I and Q inputs need to be fed with $\pm\cos \omega 1$ and $\pm\sin \omega 1$ respectively as shown in Fig. 3.5.

Fig. 3.1 Direct GFSK modulation

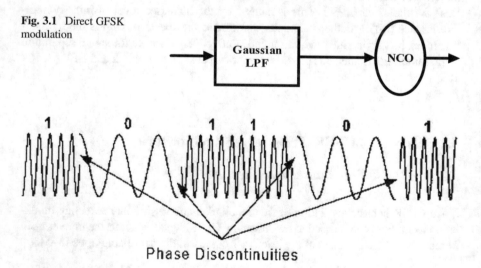

Fig. 3.2 Discontinuous phase GFSK modulation

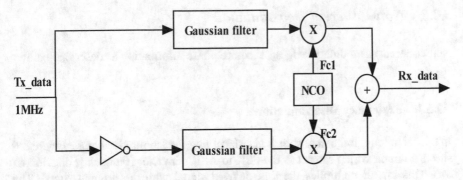

Fig. 3.3 Continuous GFSK modulation block diagram

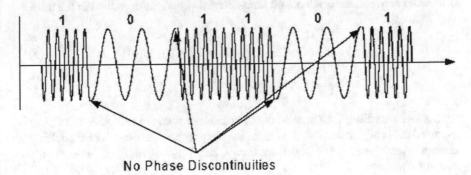

Fig. 3.4 Continuos GFSK modulation signal

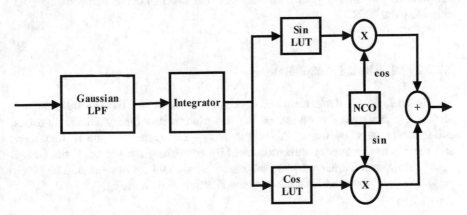

Fig. 3.5 Quadrature GFSK modulation

3.2.2 GFSK Receiver Architecture

There are two methods for Frequency demodulation (Non-coherent and coherent). They are briefly discussed in the next paragraphs [6].

3.2.3   Non-Coherent Demodulation

Non-coherent demodulation doesn't require phase information as described below.

3.2.3.1   Frequency Discrimination

In this technique, frequency shift is translated into the amplitude change. Figure 3.6 shows a circuit which can act as the FSK-to-ASK converter (Frequency discriminator). This circuit multiplies the time-delayed signal with the original signal. The output of the frequency discriminator block with time delay depends on the phase difference between the original and time-delayed signals as described in the following equations.

$$Vout = Cos(2\pi + \theta) * Cos(2\pi + \theta + \varphi(\tau)) \tag{3.1}$$

$$\frac{1}{2}Cos(\varphi(\tau)) + Cos(4\pi f + 2\theta + \varphi(\tau)) \tag{3.2}$$

If a low pass filter is used after the discriminator then the second term in (3.2) is assumed to be eliminated. So the output is solely dependent on it. The time delay is chosen in such a manner $V_{out} = 0$ for $f <= f_c$ and $V_{out} = 1$ for $f > f_c$, where fc is the center frequency.

The frequency discriminator has simple design but if there is any amplitude variation in the input signal then they get translated into the demodulated data degrading the BER.

3.2.3.2   DLL Based Demodulator

Simple DLL based demodulator is presented by Sangjin [1]. The demodulator utilizes the fact that the period of the down converted binary GFSK signal is continuously switched from $1/(f_{if} - f)$ to $1/(f_{if} + f)$, where f_{if} is the IF frequency and the f is the frequency deviation. And by comparing the period of the GFSK signal with the reference of 1/fif, the binary data could be recovered. The block diagram of the proposed DLL based demodulator is shown in Fig. 3.7.

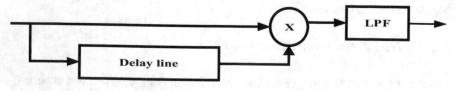

Fig. 3.6 Frequency discrimination

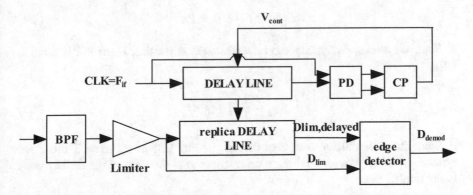

Fig. 3.7 DLL Based GFSK Demodulator

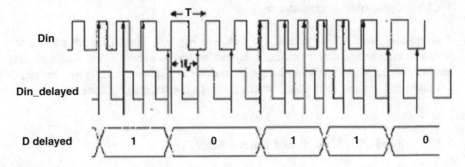

Fig. 3.8 DLL based demodulation waveforms

Since the DLL is locked to the reference clock of f_{if}, the replica delay line also provides the accurate delay of $1/f_{if}$ which is used as the reference in the period comparison. The limiter output D_{lim} passes through the replica delay line to be delayed signal D_{lim}, delayed. Then, their rising edges are compared by an edge-detecting $D_{flip\text{-}flop}$. When the delay between the two signals is $1/f_{if}$, the rising edge of D_{lim}, delayed leads or lags that of D_{lim} according to the received binary data. As a result, by detecting which rising edge comes first, the binary signal can be recovered from the GFSK modulated signal as shown in Fig. 3.8.

The down converted GFSK signal at IF can be represented as:

$$V_{if}(t) = A\cos\left(\int\limits_{-\infty}^{t} 2\pi f(\tau)d\tau\right) \quad (3.3)$$

Where $f(\tau)$ is the GFSK modulated frequency. The time interval T between two successive rising edges of D_{lim} is determined as the time duration over which the integrated phase is equal to:

$$\theta(t + T) - \theta(t) = 2\pi \tag{3.4}$$

Since the average frequency can be represented as the time interval T is equal to $1/f_{avg}$.

$$f_{avg} = \frac{1}{T} \int f(\tau)d\tau \tag{3.5}$$

Thus the demodulator compares the $1/f_{avg}$ with 1/fif every m = 3 times per symbol period to recover the binary data, where m is the ratio of the IF frequency to the symbol rate.

3.2.3.3 Correlator Demodulator

The non-coherent GFSK modulator is shown in Fig. 3.9, The BER performance of non-coherent demodulation methods is worse than coherent ones, multiply the received signal r(t) by cosine, integrate, and square. Then multiply r(t) by sine, integrate, and square. Add these two results to each other and pick the largest.

3.2.3.4 Band-Pass Filter Based Demodulator

Another non-coherent demodulator is shown in Fig. 3.10.where the band-pass filters is used instead of the correlators.

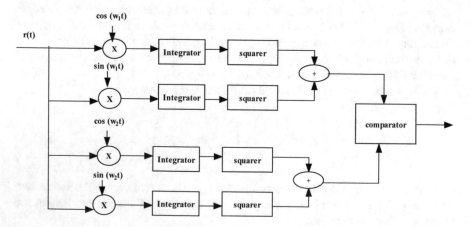

Fig. 3.9 GFSK non-coherent demodulators (correlator implementation)

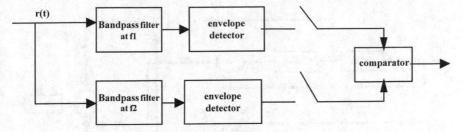

Fig. 3.10 GFSK non-coherent demodulators (band-pass filter implementation)

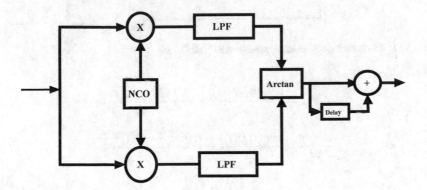

Fig. 3.11 Phase shift discrimination

3.2.4 Coherent Demodulation

Coherent demodulation requires phase information as described below.

3.2.4.1 Phase Shift Discrimination

(a) **Phase Discriminator Using (Arctan Then Differentiator)**

The Phase shift discriminator is a better demodulation method than the FSK discriminator as it utilizes only the phase of the signal Fig. 3.11 shows a phase shift discriminator. First the incoming signal is down converted into a complex Baseband signal. The two path In-phase (I) and Quadrature (Q) path are low passed filtered to eliminate the high frequency products generated by mixing. Then using an arctan block phase is extracted. In order to retrieve the NRZ signal, the output of the arctan block has to be differentiated.

(b) **Phase Discriminator Using (Differentiator of Arctan)**

There is another method for phase discriminator method without using arctan function to demodulate GFSK signals and it can derived from the following equations:

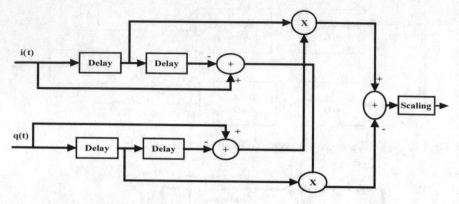

Fig. 3.12 Phase shift discriminator without ARCTAN block

$$\Delta\theta(t) = \frac{d\{\tan^{-1}[r(t)]\}}{dt} = \frac{1}{1+r^2(t)}\frac{d[r(t)]}{dt} \tag{3.6}$$

$$\frac{d[r(t)]}{dt} = \frac{d[q(t)/i(t)]}{dt} = \frac{i(t)\frac{d[q(t)]}{dt} - q(t)\frac{d[i(t)]}{dt}}{i^2(t)} \tag{3.7}$$

$$r(t) = \frac{q(t)}{i(t)} \tag{3.8}$$

$$\Delta\theta(t) = \frac{i(t)\frac{d[q(t)]}{dt} - q(t)\frac{d[i(t)]}{dt}}{i^2(t) + q^2(t)} \tag{3.9}$$

The result shown in Eq. 3.6 leads to the block diagram for the previous equations is shown in Fig. 3.12.

3.3   The Proposed GFSK Transceiver Architecture

3.3.1   The Proposed GFSK Transmitter

A block diagram of proposed GFSK is shown in Fig. 3.13. This design was preferred over other schemes discussed above because of its simplicity and continues phase as discussed before. The Numerical controlled oscillator was designed using CORDIC theory eliminating need for look up tables (save area).For the first step of modulation, No encoding schemes are used in GFSK modulation, The binary stream is used directly in the next step of modulation which is pulse shaping to reduce bandwidth. Pulse shaping is performed with a Gaussian filter (hence the name Gaussian Frequency Shift Keying). The GFSK signal is mixed with two different modulation frequencies F_1, F_2.

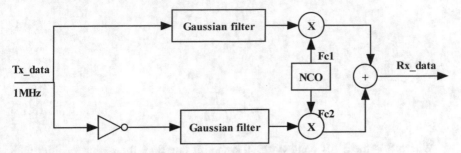

Fig. 3.13 GFSK transmitter

A GFSK signal can be written as:

$$S(t) = A \cos\left(2\pi f_c t + \phi(t, \alpha)\right) \tag{3.10}$$

Where f_c is the carrier frequency, and $\phi(t, \alpha)$ is the phase deviation, function of the input data sequence .and can be written as:

$$\phi(t, \alpha) = 2\pi h \sum_{i=n-l-1}^{n-1} \alpha_i q(t - iT) + \pi h \sum_{i=-\infty}^{n-L} \alpha_i + 2\pi h \alpha_i q[t - (n - 1)T] \tag{3.11}$$

Where α are the data bits, L is the pulse length, and for the Bluetooth standard L = 2.

The modulation index h given by:

$$h = \frac{1}{2} f_d T \tag{3.12}$$

where f_d is the frequency deviation, T the symbol time .The first sum represents symbol's contribute of L − 1 symbols that influences carrier phase deviation; the second sum represents the accumulated phase of all previous symbols that doesn't influence carrier deviation more; the third term represents the actual symbol that influence the carrier deviation The phase function q(t) is expressed as:

$$q(t) = \int_{-\infty}^{t} g(\tau) d\tau \tag{3.13}$$

Where g(t) is the output of Gaussian low Pass Filter (GLPF) for a NRZ data signal as an input. The impulse response of the GLPF is:

$$h(t) = \sqrt{\frac{\pi}{\alpha}} \exp\left(\frac{-\pi}{\alpha}t\right)^2 \tag{3.14}$$

Where

$$\alpha = \sqrt{\frac{\ln 2}{2\beta}} \tag{3.15}$$

Where β is the 3 dB bandwidth, T is the bit duration, the output of filter with rectangular pulse is:

$$g(t) = h(t) * rect\left(\frac{t}{T}\right) \tag{3.16}$$

Resolving we obtain:

$$g(t) = \frac{1}{2T}\left(Q(2\pi BT\beta_1) - Q(2\pi BT\beta_2)\right) \tag{3.17}$$

Where

$$\beta_1 = \frac{t/T - 0.5}{\sqrt{\ln(2)}} \tag{3.18}$$

$$\beta_2 = \frac{t/T + 0.5}{\sqrt{\ln(2)}} \tag{3.19}$$

And Q is a Marcum function, defined as:

$$Q(t) = \int_t^\infty \frac{1}{\sqrt{2\pi}} \exp\left(-\tau^2/2\right) d\tau \tag{3.20}$$

The impulse response of the Gaussian filter described in Eq. 3.10 can be shown in Fig. 3.14.

The mapping rule for the GFSK signals is shown in Table 3.1 where '1' is represented by positive frequency deviation and '0' is represented by negative frequency deviation.

Time signals before and after Gaussian filter is shown in Fig. 3.15.

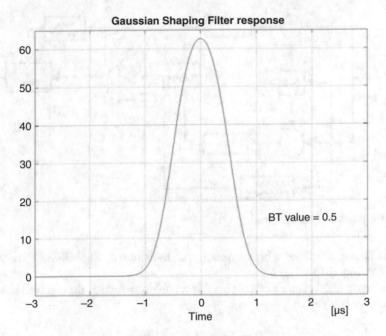

Fig. 3.14 The impulse response of the Gaussian pulse shaping filter

Table 3.1 Mapping rules for GFSK modulation

Symbol	Bit	Expression
S1	0	$\cos(2\pi(f_c - \Delta f))t$
S2	1	$\cos(2\pi(f_c + \Delta f))t$

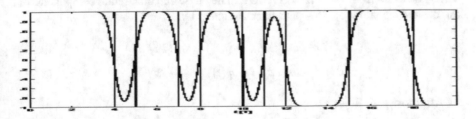

Fig. 3.15 Time signal before and after the Gaussian filter

3.3.2 The Proposed GFSK Receiver

The phase discriminator structure was chosen over other schemes discussed above because it is suitable for both GFSK and PSK modem to save area. The demodulation process is the mixing with the same modulation carrier frequency as used in the transmitter part, after the mixing the signal is low pass filtered to obtain sine and cosine components of the phase and then we get the ARCTAN of the upper and lower branches, it is differentiated and then sampled and enter the decision device

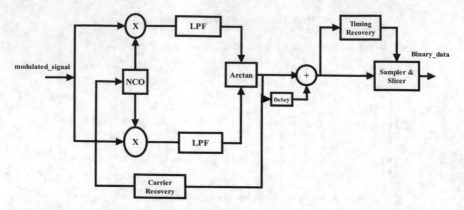

Fig. 3.16 Proposed GFSK receiver

(SLICER) to determine which symbol was transmitted, The GFSK Receiver is Shown in Fig. 3.16.

The analytical treatment of the signal received is as follow, the received signal at the receiver is given by:

$$S(t) = A \cos\left(2\pi f_c t + \phi(t, \alpha)\right) \tag{3.21}$$

where the two antipodal transmitted signals are symmetric around the carrier frequency. It is called a "mark" if the transmitted signal at the frequency $(fc + fd)$ and a "space" if it is at the frequency $(fc - fd)$. The Bluetooth standard explains that frequency deviation positive for a "mark" signal and negative for a "space" signal can be included from 140 *KHz* to 170 *KHz*. The phase and quadrature components of the desired signal before low path filter are given by:

$$x(t) = \cos\left(2\pi f_c t + \phi(t)\right) * \sin\left(2\pi f_c t\right) \tag{3.22}$$
$$x(t) = \cos\left(2\pi f_c t + \phi(t)\right) * \cos\left(2\pi f_c t\right) \tag{3.23}$$

After low path filter:

$$x'(t) = \cos\left(\phi(t)\right) \tag{3.24}$$
$$y'(t) = \sin\left(\phi(t)\right) \tag{3.25}$$

The phase of the signal is expressed as:

$$\gamma = \arctan\left(\frac{y'}{x'}\right) \tag{3.26}$$

Differentiating the above equation, we obtain the frequency deviation output; its sign determines whether it was a '1' or '0.

3.3.2.1 Proposed Timing Recovery Algorithm

The simplicity of the early-late gate algorithm made it a very good choice compared to other algorithms. While the other algorithms claim to have faster response time or estimation accuracy, these algorithms use more resources. The tradeoff between resources and performance led to the determination that the early-late gate algorithm is the preferred choice. And as mentioned before it consists of timing error detector (based on early-late gate method), second order loop filter, and numerical controlled clock (NCC). The block diagram of the proposed timing recovery algorithm is shown in Fig. 3.17.

3.4 Simulink Model for GFSK Transceiver

The first step in building the GFSK transceiver is to show how it operates mathematically. The transceiver is modeled in MATLAB (Simulink) and it consists of a GFSK transmitter, an additive white Gaussian noise (AWGN) channel (As the Bluetooth communication system is designed for radio communication in short distance and low device speed, therefore in this analyses a schemed channel by an Additive White Gaussian Noise (AWGN) is used), and a receiver.

The goal of this research is to design a 1 Mbps GFSK transceiver that meets the required Bluetooth 1.1 specifications. A simulink model for the GFSK transceiver was built to test its performance and to see if it satisfies the required specifications. Figure 3.18 shows the block diagram of the Bluetooth 1.1 transceiver simulation model adopted in this thesis. The simulation was performed at baseband taking 16 samples per symbol. Noise was added as a white Gaussian noise. In addition, phase offset (timing errors), frequency offset and drift in local oscillator frequency with time were added. A detailed description of the simulation procedure is given in the following section. The Bluetooth specification allows for L2CAP Packets to be between 0 and 65,535 bytes so the simulation should run to this long.

Fig. 3.17 Proposed timing recovery algorithm

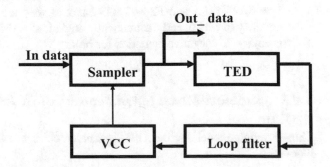

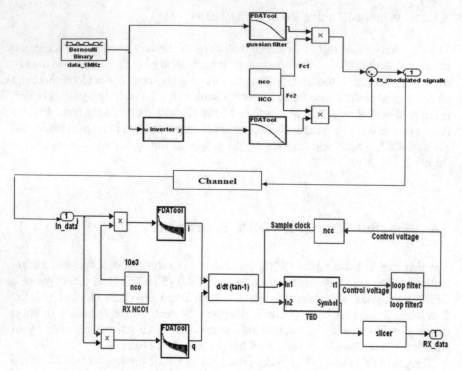

Fig. 3.18 Simulink model for GFSK transceiver

3.4.1   Simulink Model for GFSK Transmitter

The Simulink model for the GFSK transmitter consists of the following building blocks:

(a) A binary data source which generates random data at a bit rate of 1 MHz.
(b) An inverter which is the heart of continuous phase GFSK demodulation.
(c) A Gaussian filter which has Sampling frequency of 16 MHz, 65 taps, and a cutoff frequency of 0.5 MHz
(d) A numerically controlled oscillator: the frequencies were chosen to be F1 = 8.175 MHz and F2 = 7.825 MHz as they are fixed around the center frequency (8 MHz), and the frequency deviation = 175 KHz, positive deviation for binary '1', negative deviation for binary '0'.

3.4.2   Simulink Model for Bluetooth GFSK Receiver

The Simulink model for the GFSK receiver consists of the following building blocks:

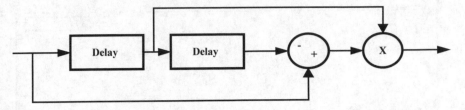

Fig. 3.19 Timing error detectors (early-late gate)

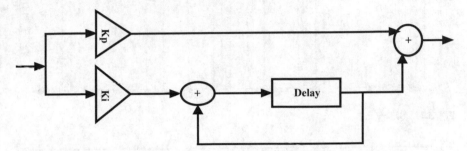

Fig. 3.20 Second order loop filter

(a) A numerical controlled oscillator (NCO) with a center frequency, Fc of 8 MHz.
(b) Low pass filter which a sampling frequency of 16 MHz and a cutoff frequency of 250 kHz.
(c) A block performing the differentiation of ARCTAN function.
(d) Sampler and slicer with Sampler clock of 1 MHz.
(e) Timing recovery which consists of a numerically controlled clock (NCC), the timing error detector based on early-late gate method and is shown in Fig. 3.19, and the second order loop filter is shown in Fig. 3.20.

The input to the timing error detector splits into three branches: one which has no delay, one which has one cycle of delay, and the final has two cycles of delay. The slope of the input is determined by subtracting the top branch from the lower branch. If the slope is zero, the clock is locked .The center branch is multiplied by the slope to determine if the slope is positive or negative. Thereby, the algorithm will sample either earlier or later until the ideal sampling time is determined. If the derivative of the input is flat, the ideal sampling time has been reached.

3.4.3   Simulink Simulation Results for GFSK Transceiver

During simulation of the GFSK transceiver, the following parameters were taken into consideration as impairments:

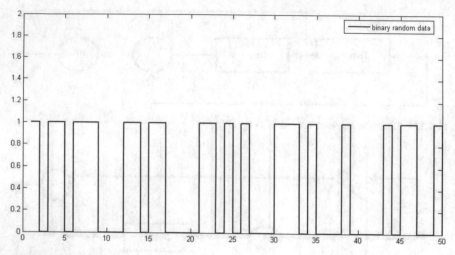

Fig. 3.21 Binary data

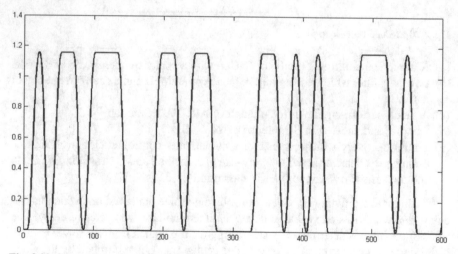

Fig. 3.22 Binary data after Gaussian filter

(a) Frequency offset, drift in local oscillator with time.
(b) Phase offset (timing error).
(c) AWGN channel.

A generic parameterized model was created with Simulink. The following Figures Shows the symbols to be transmitted using GFSK Transmitter, at the previous conditions in Fig. 3.21 and after Gaussian filtered in Fig. 3.22, These symbols after being multiplicities with the carrier frequency as shown Figs. 3.23 and 3.24 and the transmitted GFSK modulated signal as shown in Fig. 3.25, Received I & Q components as shown in Fig. 3.26, and The received symbols in Fig. 3.27. It is apparent that timing recovery is a very important issue in designing of GFSK transceiver as it improves SNR.

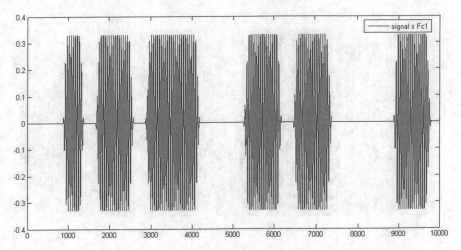

Fig. 3.23 In-phase components

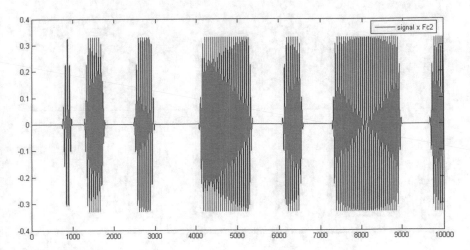

Fig. 3.24 Quadrature component

3.5 HDL Model for GFSK Transceiver

The previous simulation shows that the GFSK transceiver is capable of communicating a message between the transmitter and the receiver mathematically, and in simulation. However the transceiver needs to be built in hardware. The design is mapped into Xilinx Spartan3 device using (FPGA Advantage and Xilinx ISE6.3 tools), Synthesis & optimization were done using the LEONARDO SPECTRUM Synthesizer. In the following sections we will discuss the VHDL model for the GFSK transceiver, its algorithms, and architectures [7–9].

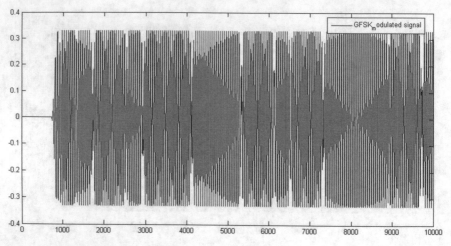

Fig. 3.25 Transmitted GFSK signal

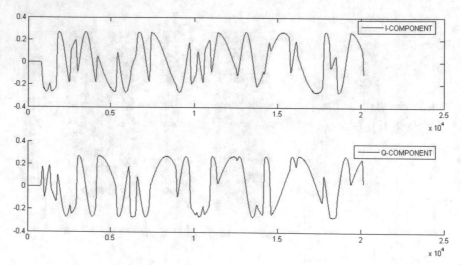

Fig. 3.26 Received I & Q components

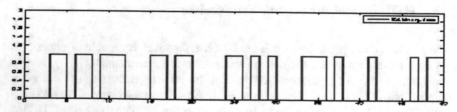

Fig. 3.27 Received binary data

3.5.1 HDL Model for GFSK Transmitter

The HDL model for the GFSK Transmitter is shown in Fig. 3.28 which resembles that of the GFSK modulator Bluetooth 1.1 Core Radio.

The main pins description for the top level entity of the GFSK modulator is summarized in Table 3.2.

It may be noticed that GFSK transmitter consists of inverter, two Gaussian filters (pulse shaping), numerically controlled oscillator (NCO) which generate two frequencies (F_1 to represent logic '1' and F_0 to represent logic '0'), two multipliers and adder. The critical components in this design are the numerically controlled oscillator and the Gaussian filter and they will be discussed in details in next section.

3.5.1.1 HDL Model for the Numerically Controlled Oscillator (NCO)

The VHDL model for the numerically controlled oscillator can be shown in Fig. 3.29. The design is based on CORDIC theory (rotation mode) as it more area effective than using look-up tables for sine and cosine. It consists of a phase accumulator, first quadrant adjustment; a pipelined unrolled CORDIC which

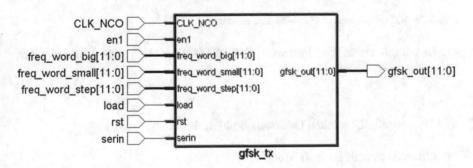

Fig. 3.28 HDL model for GFSK transmitter

Table 3.2 Pin description of VHDL model of GFSK modulator

Pin name	Description
Clk_NCO	Sampling frequency
load	To load value into NCO to generate desired carrier wave
rst	Enable high
Serin	Serial data at bit rate = 1 Mbps
En1	Enable high
Freq_word_step	To adjust phase offset of the carrier signal
Freq_word_big	To generate f_c corresponding to '1'
Freq_word_small	To generate f_c corresponding to '0
GFSK_out	GFSK modulated signal

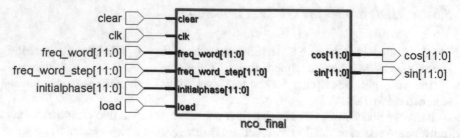

Fig. 3.29 VHDL model for the numerically controlled oscillator

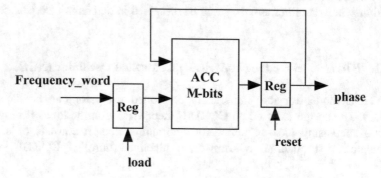

Fig. 3.30 RTL schematics of the phase accumulator

consists of subsystem of 13 iterations of a CORDIC base block, sine-cosine rebuilds, and delay sub-modules.

The Numerically Controlled Oscillator Building Blocks

(a) Phase Accumulator Sub-Module:

The phase accumulator sub-module is an n-bit accumulator whose value specifies the instantaneous phase as shown in Fig. 3.30.

The M-bit value "frequency control word" fed as the argument ().At each clock cycle (clock frequency), the phase accumulator increments itself by the value until it overflows and wraps around. Each overflow of the accumulator corresponds to one period of a sine (or cosine) wave. Thus, controls the rate at which the accumulator overflows, thereby controlling the frequency of the sine or cosine waveform. The value of the M -bit accumulator represents a signed fractional number in the interval $[-1, 1]$. This value is mapped to an angle (or phase value) in the interval $[-\pi, \pi]$. The output frequency can be calculated as [10]:

Table 3.3 To first quadrant mapping rules

Angle	Normalized angle	Correspond integer value	Binary equivalent
0	0	0	000000000000
$\pi/8$	1/8	256	000100000000
$\pi/4$	2/8	512	001000000000
$3\pi/8$	3/833 3/8	768	001100000000
$\pi/2$	4/8	1024	010000000000
$\pi/2+\pi/8$	5/8	1280	010100000000
$3\pi/4$	6/8	1536	011000000000
$7\pi/8$	7/8	1792	011100000000
π	1	2047	011111111111
$\pi+\pi/8$	9/8	2304	100100000000
$5\pi/4$	10/8	2560	101000000000
$11\pi/8$	11/8	2816	101100000000
$3\pi/2$	12/8	3072	110000000000
$13\pi/8$	13/8	3328	110100000000
$14\pi/8$	14/8	3584	111000000000
$15\pi/8$	15/8	3840	111100000000
$-\pi$	-1	4096	100000000000

$$\text{Fcw} = \Delta = (F_{out}/F_{clk})2^M. \tag{3.27}$$

According to Nyquist theorem, since at least two samples are required per clock cycle, the largest value to which Δ can be assigned is $2^{\wedge M-1}$, and therefore the maximum output frequency is limited to $F_{clk}/2$.

(b) **First Quadrant Adjustment Sub-Module**

Because the CORDIC algorithm works in the first and fourth quadrants only so a block is needed to Map values of () from (0:2π) to (0: π/2), the two most significant bits of all the quadrants are set to "00",.Mapping to first quadrature example is shown in Table 3.3.

(c) **CORDIC Pipelined Sub-Module**

This module calculates the value of sin () & cos () in the range to (0: π/2) using the CORDIC algorithm, and lookup values of ARCTAN which has depicted in Table 3.4.

The top level VHDL model for the CORDIC pipelined sub-module is shown in Fig. 3.31, it consist of 13 consecutive instances of the CORDIC-base sub-block. The number of iterations was chosen as 13 as it gives the best accuracy in calculation of sine and cosine functions.

Table 3.4 Lookup table for ARCTAN

Iteration	Value	Iteration	Value
1	45	8	0.4476
2	26.5651	9	0.2238
3	14.0362	10	0.1119
4	7.1250	11	0.0560
5	3.5763	12	0.0280
6	1.7899	13	0.0140
7	0.8952		

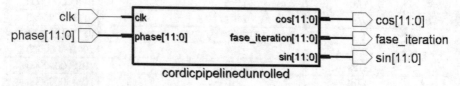

Fig. 3.31 VHDL model for pipelined CORDIC

Table 3.5 De-mapping of sine and cosine values

MSB	Sin	Cos
00	sin φ	cos φ
01	cos φ	−sin φ
10	−sin φ	−cos φ
11	−cos φ	sin φ

(d) Sine Cosine Rebuild Sub-Module

This module de-maps the values of sin () and cos () to the range (0:2π). The de-mapping algorithm is shown in Table 3.5.

(e) Delay Block Sub-Module

A delay is required for the two most significant bits of the angle to control the sine cosine rebuild block (synchronization).

HDL Simulation Results for NCO

In this example of simulation: $F_{out}/F_{clk} = 1/16$, $1/32$, $1/64$ as shown in Fig. 3.32.

3.5.1.2   HDL Model for Gaussian Filter

The general architecture of the finite impulse response filter is shown in Fig. 3.33.
The following equation represents an FIR filter operation:

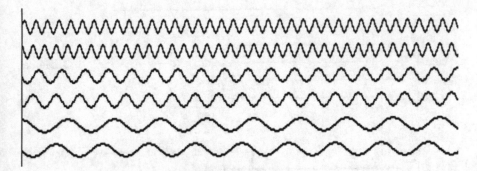

Fig. 3.32 Sine-cosine HDL simulation results

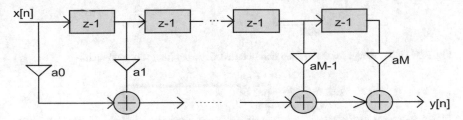

Fig. 3.33 General architecture of FIR digital filter

$$y(n) = \sum_{i=0}^{L-1} x(n-i)h(i) \tag{3.28}$$

Where x(n) represents the sequence of input samples, h(n) represents the filter coefficient L is the number of filter taps.

Because the data is serial and represented by one bit there is no need for multipliers,it we use multiplexers instead, that is, If input data is '1' the output data equals the coefficient, otherwise the output is zero. The Gaussian filter was hand-crafted and has 65 taps (16 MHz *4 + 1) where the filter period is $[-4T, 4T]$, The VHDL model for 1 tap of the Gaussian filter is shown in Fig. 3.34. The coefficient of the filter was generated using filter design and analysis tool in MATLAB (FDATOOL).

3.5.2 HDL Model for GFSK Receiver

The HDL model for the GFSK receiver is shown in Fig. 3.35 which resembles that of the GFSK modulator of Bluetooth 1.1 Core Radio. The main pins description for the top level entity of the GFSK modulator is summarized in Table 3.6.

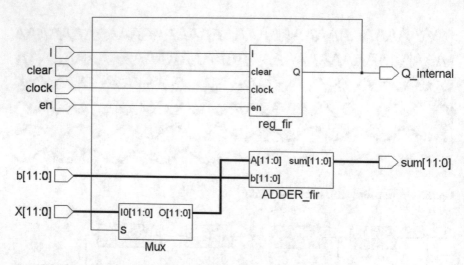

Fig. 3.34 HDL model for the one tap hand-crafted Gaussian filter (no multipliers)

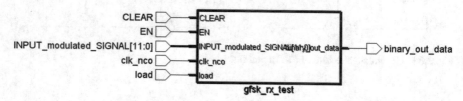

Fig. 3.35 HDL model for GFSK receiver

Table 3.6 Pin configuration for GFSK receiver

Pin name	Description
Clk_sampling	Sampling frequency
load	To load value into NCO to generate desired carrier wave
Clear	Enable high
Input_modulatd_signal	Modulated GFSK signal
EN	Enable high
Freq_word_step	To adjust phase offset of the carrier signal
Clk_bit	For the slicer
Binary_out_data	The binary data after demodulation

It is apparent that the GFSK receiver consists of multipliers, two low pass filters, numerically controlled oscillator (NCO) which generates center carrier frequency, ARCTAN block, differentiator, sampler, decision maker, and timing recovery. The critical components in this design is the ARCTAN function low pass filter and timing recovery .These components will be discussed in details in next section.

3.5.2.1   HDL Model for ARCTAN Function

The ARCTAN block is implemented using the CORDIC Algorithm as previously mentioned in Chap. 1; CORDIC algorithm has two modes of operation: Rotation mode (to generate sine and cosine functions) and vectoring mode (to calculate ARCTAN function.

The Rotation mode CORDIC Algorithm can be summarized by the following equations:

$$X_{i+1} = X_i - S_i 2^{(-2i)} Y_i \tag{3.29}$$

$$y_{i+1} = Y_i + S_i 2^{(-2i)} X_i \tag{3.30}$$

$$\Phi_{i+1} = \Phi i - S_i \arctan 2^{(-2i)} \tag{3.31}$$

Where

$$Si = 1 \text{ if } \Phi_i > 0 \text{ else} - 1 \tag{3.32}$$

Where (X_i, Y_i) is a vector and (X_{i+1}, Y_{i+1}) is (X_i, Y_i) vector after rotation by Φ angle.

Figure 3.36 summarize the above equations.

The Vectoring mode CORDIC Algorithm differs from rotation mode algorithm in that direction of rotation is determined by the sign of y instead of Φ ($Si = 1$ if $yi < 0$ else -1) as shown in Fig. 3.37.

Fig. 3.36 Rotation mode CORDIC algorithm

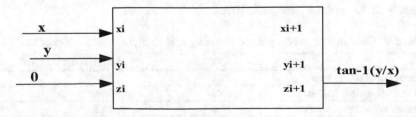

Fig. 3.37 Vectoring mode CORDIC algorithm

HDL Simulation Results for ARCTAN Block

HDL Simulation Results for ARCTAN are shown in Fig. 3.38, where in this example x = y = 1 degree so ARCTAN (y/x) =45°.

Another example is shown in Fig. 3.39, where ARCTAN (0.5) =63.5^0 (normalized values).

The proposed thesis analyzed and implemented the phase discriminator by two methods. It turns out the first method (using ARCTAN then differentiator) consumes more area than the second method (differentiator of ARCRAN). Table 3.7 shows this comparison.

The first method was chosen for the proposed architecture e because it can be used for both GFSK and DPSK transceiver.

3.5.2.2   HDL Model for Low Pass Filter

The coefficients and the VHDL code were generated using FDATOOL in matlab7 (65 Taps filter). The design was generated using the windowing method (sampling frequency is 16 MHz and the cutoff frequency is 0.5 MHz).The FDATOOL method was preferred over hand –crafted filter method because the outcome consumes less area. It was preferred over using Xilinx core generator because it is technology–dependent (maps only into Xilinx chips). Table 3.8 shows a comparison between designing the FIR filter using hand-crafted method, FDATOOL method and Xilinx core generator. Synthesis results refer to mapping into a spartan3 (200 g256) chip.

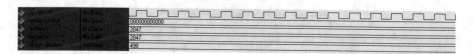

Fig. 3.38 HDL simulation Result for ARCTAN

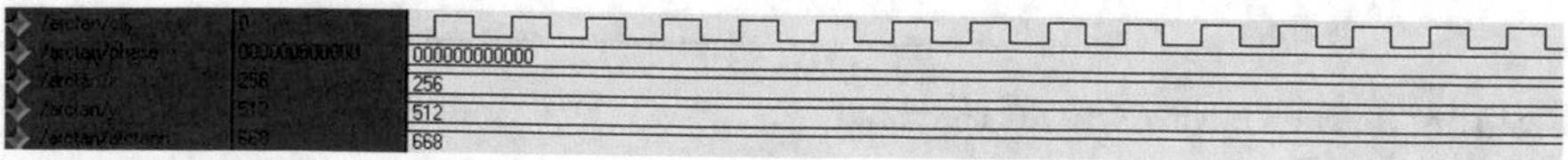

Fig. 3.39 HDL simulation result for ARCTAN

Table 3.7 Comparison between synthesis results for ARCTAN and /Differentiator of ARCTAN

	ARCTAN	Differentiator of ARCTAN
Number of slice flip flops:	12%	2%
Number of 4 input LUTs:	13%	6%
Number of occupied slices:	15%	10%
Number of MULT18X18s:	–	16%

Table 3.8 Comparison of different methods of implementation FIR filter

	Hand-crafted	FDATOOL	Core generator
Number of 4 input LUTS	6%	2.5%	2%

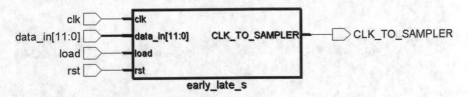

Fig. 3.40 Top level entity for timing recovery

3.5.2.3 HDL Model for Timing Recovery

The timing recovery block consists of timing error detector (based on early-late gate method), second order loop filter (to filter the timing error in order to provide the required correction to the numerical controlled clock, and a numerically controlled clock (NCC). This is implemented in the same way as the NCO except that it generates square wave rather than a sinusoidal. The top level of the VHDL entity model of timing recovery is shown in Fig. 3.40 [11].

3.5.3 HDL Simulation Results for GFSK Transceiver

Figure 3.41 shows the simulation results of GFSK transceiver, where the received signal are the same as the transmitted one.

3.6 Conclusion

In this chapter, GFSK modulation-demodulation is discussed. SIMULINK model for it was built and VHDL model too, Timing recovery technique is used which improves BER performance of the transceiver.

References

1. H. Ishikuro, M. Hamada, A single-chip CMOS Bluetooth transceiver with 1.5MHz IF and direct modulation transmitter. In *IEEE Int. Solid-State Circuits Conf*, Feb 2003, pp. 272–273
2. S. Mukthavaram, *Design and FPGA Implementation of an Adaptive Demodulator* (B.S. E.E Osmania University, Hyderabad, India, 1997)

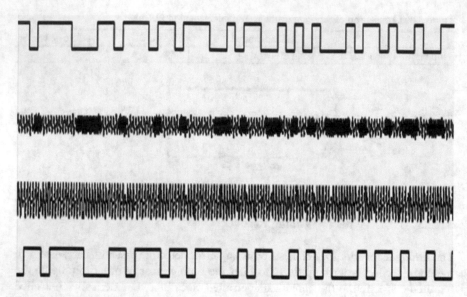

Fig. 3.41 VHDL simulation results for GFSK transceiver

3. H. Meyr, M. Moeneclaey, S.A. Fechtel, *Digital Communication Receivers: Synchronization, Channel Estimation, and Signal Processing* (Wiley, New York, 1998)
4. J.A.C. Bingham, *The Theory and Practice of MODEM Design* (Wiley, New York, 1988)
5. R. Schiphorst, F. Hoeksema, K. Slump, *Bluetooth Demodulation Algorithms and their Performance* (University of Twente, Netherlands, 2003)
6. D.S. Gomez, *Analysis of a Software Bluetooth Modem based on a DSP Implementation* (Atizapn de Zaragoza, Estado de México, 2004)
7. Hardware Description Language. Wikipedia: The Free Encyclopedia. 18 May 2004. http://en.wikipedia.org/wiki/Hardware_description_language
8. S. Yalamanchili, *Introductory VHDL, From Simulation to Synthesis* (Prentice Hall, Netherlands, 2001)
9. U. Meyer-Baese, *Digital Signal Processing with Field Programmable Gate Arrays* (Springer, Netherlands, 2001)
10. A.G. Bogdan, K H. Joshua, et al., A 0.18um CMOS Bluetooth frequency synthesizer for integration with a Bluetooth SOC reference platform. In *IEEE Int. System-on-Chip*, 2003
11. www.mathworks.com
12. F. Li, R. Butaud, E. Dekneuvel, G. Jacquemod, *Bluetooth Transceiver Modeling Using System C-AMS*. IEEE PRIME, Villach, Austria, 2013
13. M.S. Pereira, J.C. Vaz, C.A. Leme, J.T. Sousa, J.C. Freire, A 170 μA all-digital GFSK demodulator with rejection of low SNR packets for Bluetooth-LE. IEEE Microw. Wirel. Compon. Lett. **26**(6), 452–454 (2016). https://doi.org/10.1109/LMWC.2016.2562639
14. S. Gao, H. Jiang, Z. Weng, Y. Guo, J. Dong, et al., A 7.9 μA multi-step phase-domain ADC for GFSK demodulators. Analog Integr. Circ. Sig. Process **94**(1), 49–63 (2018). https://doi.org/10.1007/s10470-017-1081-5

15. B. Chi, J. Yao, P. Chiang, Z. Wang, A 0.18-um CMOS GFSK analog front end using a Bessel-based quadrature discriminator with on-chip automatic tuning. IEEE Trans. Circuits Syst. I Regul. Pap. **56**(11), 2498–2510 (2009). https://doi.org/10.1109/TCSI.2009.2015728
16. D.S. Chirov, A.N. Vynogradov, E.O. Vorobyova, Application of the decision trees to recognize the types of digital modulation of radio signals in cognitive systems of HF communication. In *Proc. Syst. Signal Synchronization, Generating Process. Telecommun. (SYNCHROINFO)*, Jul. 2018, pp. 1–6
17. A. Latif, N.D. Gohar, Error rate performance of hybrid QAM-FSK in OFDM systems exhibiting low PAPR. Sci. China F, Inf. Sci. **52**(10), 1875–1880 (2009)
18. M. Pischella, D. Le Ruyet, *Digital Communications 2: Digital Modulations*, 1st edn. (ISTE, London, UK, 2015) ISBN 978-18-4821-846-8

Chapter 4
Hardware Realization of DPSK-Based Bluetooth Modem

4.1 DPSK Transceiver Overview

π/4-DQPSK has no 180° phase shifts as the maximum phase transition is limited to 135° thereby limiting spectral growth, This allows us to place channels closer to each other thereby maximizing the limited spectrum of the ISM band. In addition, it can be differentially demodulated. These properties make it particularly suitable for mobile communication, where differential modulation can reduce the adversary effect of the fading channel [1].

In fact, every data packet always starts with an Access Code and a Header Code modulated by GFSK and if needed the Payload can be transmitted with the two new modulation types to improve the data rate. This feature has been chosen in order to guarantee an entire compatibility with the basic GFSK standard, but in this way, the resulting packet structure is more complicated. Data transmission through the new modulation types needs a Guard Time to separate the two kinds of modulations [2–13].

4.2 Proposed DPSK Transceiver Architecture

The proposed architecture of the DPSK transmitter and receiver are detailed in this section.

4.2.1 Proposed DPSK Transmitter Architecture

The structure of a typical digital DPSK transmitter is shown in Fig. 4.1. The transmitter consists of two branches: one for the In-phase (I) channel and one for

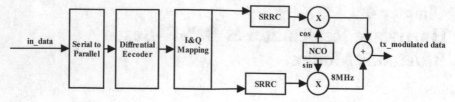

Fig. 4.1 DPSK transmitter

Fig. 4.2 Differential
encoding

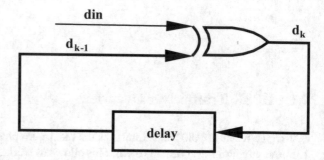

the quadrature (Q) channel. The operation of the transmitter can be understood by tracing the flow of data through the functional blocks inside the transmitter [14–16].

This modulator is the same for both π/4-DQPSK and 8-DPSK except for the serial to parallel clock (2 MHz for π/4-DQPSK, 3 MHz for 8DPSK) and in in-phase quadrature mapping rules. The Serial to Parallel block converts incoming serial data into two $N/2$ bit words per symbol for π/4-DQPSK and $N/3$ bit words per symbol for 8DPSK and Hence, the symbol rate is $1/N$ times the bit rate. The data is then differentially encoded at the symbol rate and then fed to in-phase and quadrature phase mapper. The mapped values are subsequently filtered by SRRC filters to limit the bandwidth of the transmitted signal without introducing Inter-Symbol Interference (ISI) as described in Chap. 1. A Square Root Raised Cosine filter (SRRC) with a roll-off factor α equal to 0.4 is required by Enhanced Data Rate specifications, both transmission and receiving filter.

Differential encoding is used to provide polarity reversal protection. Bit streams going through the many communications circuits can be unintentionally inverted. Most signal processing circuits can not tell if the whole stream is inverted. This is also called phase ambiguity. Differential Encoding is used to protect against this possibility. It is one of the simplest forms of error protection coding done on a baseband sequence prior to modulation. A differential Coding system consists of a modulo 2 adder operation as shown below in Fig. 4.2. The decoding process reverses the above as show in Fig. 4.3 [17].

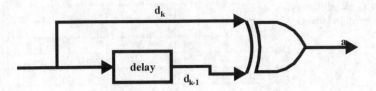

Fig. 4.3 Differential decoding

Fig. 4.4 π/4DQPSK
constellation

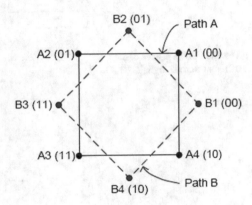

4.2.1.1 π /4-DQPSK Mapping

Similar to QPSK, π/4DQPSK transmits two bits per symbol. However, there is a subtle difference. In QPSK there are four signals that are used to send the four symbols. In π/4DQPSK there are eight signals; every alternate symbol is transmitted using a π/4 shifted pattern of the QPSK constellation [18–20].

Symbol A uses signal on Path A as shown below in Fig. 4.4 and the next symbol *B*, even if it is exactly the same bit pattern uses a signal on Path B. So we always get a phase shift even when the adjacent symbols are exactly the same.

The constellation diagram appears to be similar to 8-PSK. Note that an 8-PSK constellation can be broken into two QPSK constellations as shown below. In π/4DQPSK, one symbol is transmitted on the A constellation and the next one is transmitted using the B constellation.

The mapping rule for π/4DQPSK symbols are shown in Table 4.1.

4.2.1.2 8DPSK Mapping

8DPSK transmitted signal shows smaller phase transitions (on average) than QPSK but since the signals are also less distinctly different from each other, makes 8DPSK prone to higher bit errors. But using 8DPSK we can convey three bits per symbol. The throughput of 8DPSK is 50% better than QPSK which can transmit just 2 bits per symbol as compared to 3 for 8DPSK.

Table 4.1 π/4DQPSK symbols mapping to I and Q

bit	ID	I	Q
00	A1	0.707	0.707
00	B1	0	1
10	A4	0.707	−0.707
10	B4	−1	0
01	A2	−0.707	0.707
01	B2	1	0
11	A3	−0.707	−0.707
11	B3	0	−1

Fig. 4.5 8DPSK uses eight different unique signals

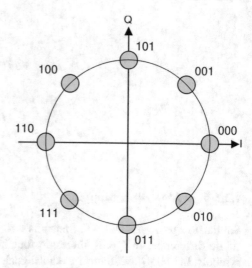

Table 4.2 8DPSK symbols mapping to I and Q

bit	Symbol	I	Q
000	S1	1	0
001	S2	0.7	0.7
011	S3	0	1
010	S4	−0.7	0.7
110	S5	−1	0
111	S6	−0.7	−0.7
101	S7	0	−1
100	S8	0.7	−0.7

8DPSK is the first of the bandwidth-efficient modulations. The constellation diagram of 8DPSK is shown in Fig. 4.5, Table 4.2 Shows Mapping rules for 8DPSK [9, 10].

After mapping the data is pulse shaped using square root raised cosine filters (SRRC) whose impulse response can be shown in Fig. 4.6 and then the signal multiplied by the carriers.

Fig. 4.6 Impulse response of the square root-raised cosine

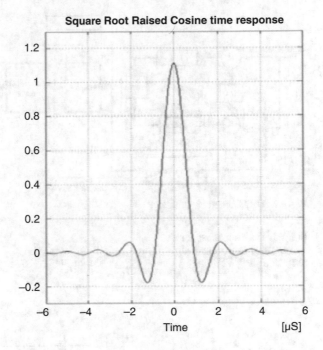

Therefore a band-pass signal is obtained and transmitted, which has the following form:

$$s(t) = A \cos \left(2\pi ft + \varphi(t) \right) \qquad (4.1)$$

Where $\phi(t)$ is carrier phase, given by:

$$\varphi(t) = \frac{2\pi(i-1)}{M} \qquad (4.2)$$

Where $i = 0$ to 3 (for $\pi/4$DQPSK) and $i = 0$ to 7 (for 8DPSK).

Using trigonometric identities the transmitted signal can be split into In-phase component I(t) and Quadrature-phase component according to:

$$s(t) = I(t) \cos \left(2\pi ft \right) + Q(t) \sin \left(2\pi ft \right) \qquad (4.3)$$

4.2.2 The Proposed DPSK Receiver Architecture

The structure of a typical digital DPSK receiver implemented in digital logic is shown in Fig. 4.7. There is no non-coherent detection equivalent for DPSK as

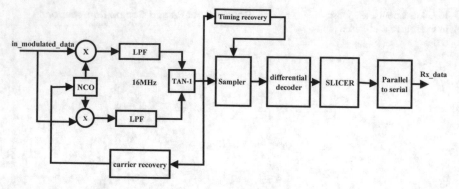

Fig. 4.7 DPSK receiver

Fig. 4.8 Decision zones for
π /4-DQPSK Demodulation
[22]

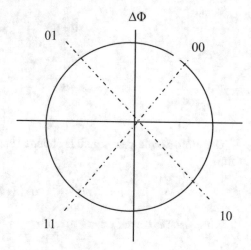

non-coherency implies no phase information. The operation of the receiver can be understood by tracing the flow of data through the functional blocks inside the receiver. The demodulation process is the same for both π /4-DQPSK and 8-DPSK except for the sampler clock, slicer (de-mapping rules), and parallel to serial clock.

The demodulation process involves the mixing with the same modulation carrier frequency as used in the transmitter part, after the mixing the signal is low pass filtered to obtain sine and cosine components of the phase and then we get the ARCTAN of the upper and lower branches, it then sampled and enter the decision device (slicer) to determine which symbol was transmitted [21].

Figures 4.8 and 4.9 show the decision zones for π /4-DQPSK and 8DPSK respectively, timing and carrier recovery are carried too; the analytic equations are the same as in GFSK except that the phase is not differentiated.

Fig. 4.9 Decision zones for 8DPSK demodulation [23]

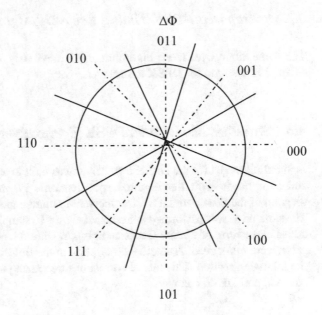

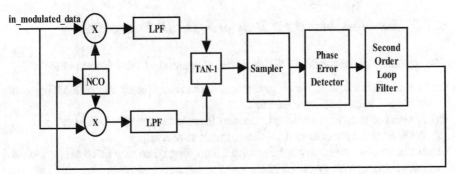

Fig. 4.10 Proposed carrier recovery

4.2.3 Proposed DPSK Carrier Recovery Algorithm

If the receiver's and transmitter's clocks are misaligned, the resulting can impair reception on either arm of the receiver. Thus, a carrier recovery algorithm is required. The proposed Carrier recovery algorithm is shown in Fig. 4.10 where the phase error is generated from the in-phase and quadrature phase branches. This scheme was preferred because it gives more accurate results; the carrier recovery algorithm consists of a numerically controlled oscillator, a phase detector and a second order loop filter [24].

4.2.4 Proposed DPSK Timing Recovery Algorithm

The same timing recovery algorithm used previously for GFSK transceiver in Chap. 2 is used too for DPSK transceiver.

4.3 Simulink Model for 8DPSK Transceiver

A Simulink model for the DPSK transceiver was built to test its overall performance and compliance with the required specifications. Figure 4.11 shows the block diagram of the Bluetooth DPSK transceiver simulation model adopted in this thesis. The simulation was performed at baseband taking 16 samples per symbol; Noise was added in the form of a white Gaussian noise. In addition, phase offset and frequency offset were introduced. A detailed description of the simulation procedure is given in the following section. This model consist of three main parts the DPSK transmitter, channel, and DPSK receiver.

4.3.1 Simulink Model for Bluetooth DPSK Transmitter

The simulink model for the DPSK transmitter consists of the following blocks:

(a) A binary data source which generate random data at a bit rate of 2 MHz for $\pi/$4DQPSK and 3 MHz for 8DPSK.
(b) A serial to parallel converter to convert bit rate to symbol rate of 1 Mbps.
(c) A Differential encoder and in-phase, quadrature mapper
(d) A square root raised cosine filter with a Sampling frequency of 16 MHz, 65 taps, and cutoff frequency of 0.5 MHz.
(e) A numerically controlled oscillator: the center frequency is chosen to be 8 MHz.

4.3.2 Simulink Model for Bluetooth DPSK Receiver

The simulink model for the DPSK receiver consists of the following building blocks:

(a) A numerical controlled oscillator (NCO) with a center frequency of 8 MHz.
(b) A low pass filter with a sampling frequency of 16 MHz and a cutoff frequency of 0.5 MHz.
(c) A block performing the ARCTAN function.
(d) A sampler and slicer with a Sampling of 1 MHz.
(e) Timing recovery which consists of a numerically controlled clock (NCC). The Timing error detector is based on the early-late gate method. The second order Loop was Filter discussed in the previous chapter.

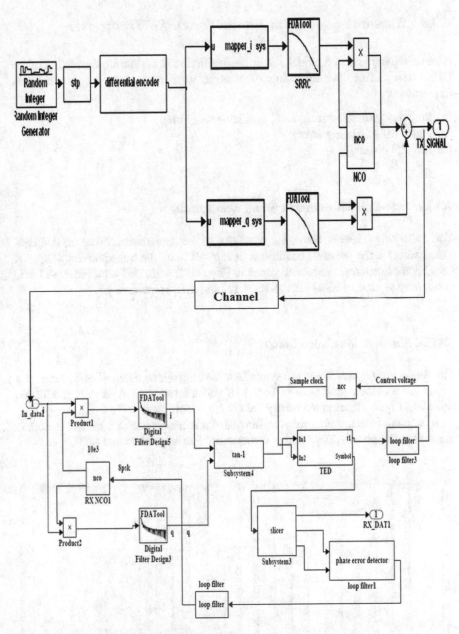

Fig. 4.11 Simulink model for DPSK transceiver

(f) Carrier recovery which consists of a numerically controlled clock, a phase error
    detector and a second order loop path filter.

4.3.3   Simulink Simulation Results for DPSK Transceiver

A generic parameterized model was created with Simulink. During simulation of the DPSK transceiver, the following parameters were taken into consideration as impairments:

(a) Frequency offset, drift in local oscillator with time.
(b) Phase offset (timing error).
(c) AWGN channel.

4.3.3.1   π/4-DQPSK Simulink Simulation Results

The following Figures show the symbols to be transmitted using **π/4DQPSK** Transmitter, at the previous conditions in Fig. 4.11 and the transmitted **π/4DQPSK** I & Q component and modulated signal in Figs. 4.12, 4.13, and 4.14, received I & Q and binary signals in Figs. 4.15, 4.16, 4.17, and 4.18 respectively.

8DPSK Simulink Simulation Results

The following Figures Show the symbols to be transmitted using 8DPSK Transmitter, at the previous conditions in Fig. 4.19 and the transmitted 8DPSK modulated signal. In Fig. 4.20, received binary data in Fig. 4.21.

It is apparent that timing recovery and carrier recovery are very critical factors in improving the phase offset (timing error) and frequency offset and SNR.

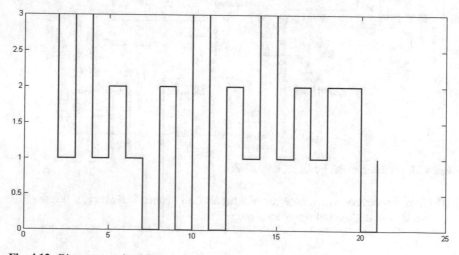

Fig. 4.12 Binary transmitted data

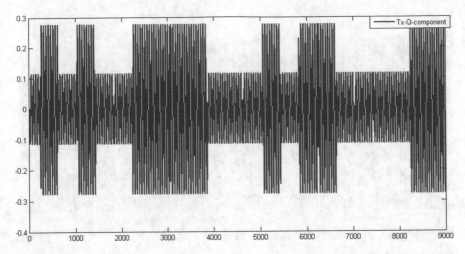

Fig. 4.13 Transmitted I-components

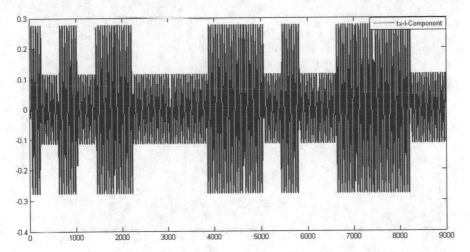

Fig. 4.14 Transmitted Q-components

4.4 VHDL Model for DPSK Transceiver

The previous simulation shows that the transceiver is capable of communicating a
message between the transmitter and the receiver mathematically, and in simulation.
However the transceiver needs to be built in hardware. The design is mapped into a
Xilinx Spartan3 device using (FPGA ADVANTAGE and Xilinx ISE 6.3 tools).
Synthesis & optimization were done using the LEONARDO SPECTRUM Synthesizer. In the following section we will discuss the VHDL model for the DPSK
transceiver, its algorithms, and architectures.

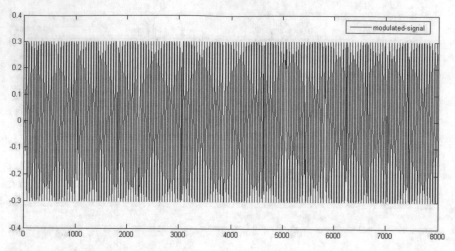

Fig. 4.15 π/4DQPSK modulated transmitted data

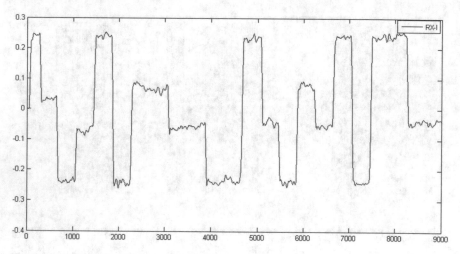

Fig. 4.16 Received I-component

4.4.1 VHDL Model for DPSK Transmitter

The VHDL model for the DPSK Transmitter shown in Fig. 4.22 resembles that of the DPSK modulator of the Bluetooth.

The main pins description for the top level entity of the DPSK modulator is summarized in Table 4.3.

It may be noticed that DPSK transmitter consists of Serial to parallel block (2Mbps for π/4DQPSKand 3Mbps for 8DPSK),in-phase and quadrature phase mapping (mapping rules differs for π/4DQPSK from 8DPSK), two square root raised cosine filters(pulse shaping),a numerically controlled oscillator (NCO)

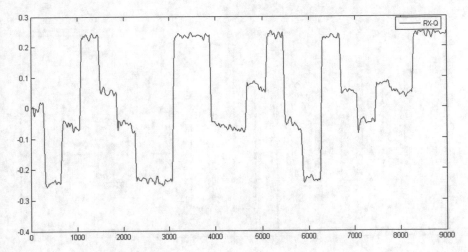

Fig. 4.17 Received Q-component

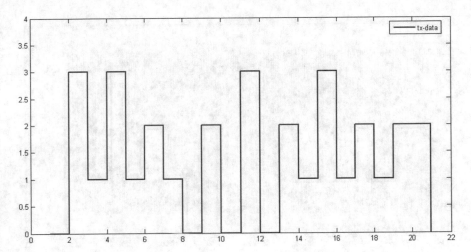

Fig. 4.18 Received binary data

which generates the carrier frequency, two multipliers and an adder as shown in figure. The critical components in this design are the numerically controlled oscillator and the square root raised cosine filter (SRRC) and they will be discussed in detail in the next section.

4.4.1.1   VHDL Model for Square Root Raised Cosine Filter

The square root raised cosine filter was implemented using Single Port Block Memory which provides good utilization of resources.

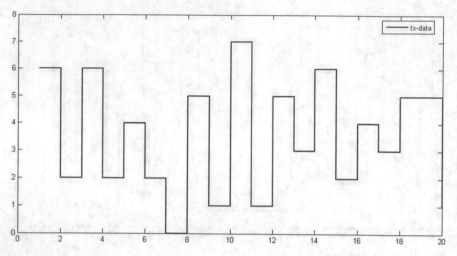

Fig. 4.19 Binary transmitted data

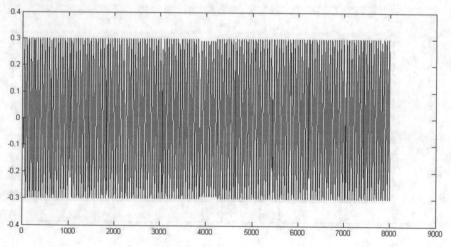

Fig. 4.20 8DPSK modulated transmitted data

4.4.2 VHDL Model for DPSK Receiver

The VHDL model for the DPSK receiver shown in Fig. 4.23 resembles that of the DPSK demodulator of Bluetooth [1].

The description of the main pins for the top level entity of the DPSK modulator are summarized in Table 4.4.

The DPSK receiver consists of multipliers, two low pass filters, a numerically controlled oscillator (NCO) which generates a center carrier frequency, an arctan block, a sampler and decision, a parallel to serial converter, timing recovery, and

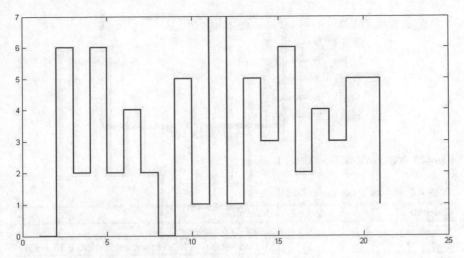

Fig. 4.21 Received binary data

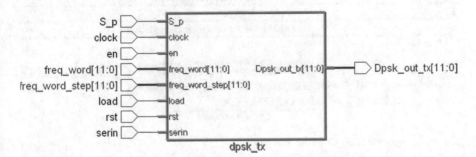

Fig. 4.22 VHDL model for DPSK transmitter

Table 4.3 Pin description of DPSK modulator

Pin name	Description
Clk_NCO	Sampling frequency
load	To load value into NCO to generate desired carrier wave
Rst	Enable high
serin	Serial data
EN	Enable high
S_p	'1' for 8DPSK, '0' for π/4DQPSK
DPSK_out_tx	Out PSK modulated signal

carrier recovery. The critical components in this design are the carrier recovery block; all other components were discussed before in Chap. 2.

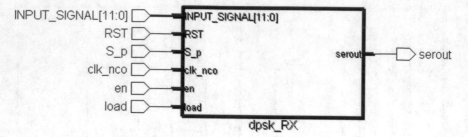

Fig. 4.23 VHDL MODEL for DPSK Receiver

Table 4.4 Pin configuration for DPSK receiver

Pin name	Description
Clk_NCO	Sampling frequency
load	To load value into NCO to generate desired carrier wave
Clear	Enable high
Input_ signal	Modulated DPSK signal
En	Enable high
Serout	The binary data after demodulation

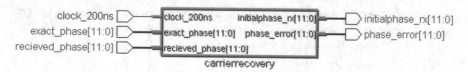

Fig. 4.24 Top level entity of carrier recovery

4.4.2.1 VHDL Model for Carrier Recovery

The phase-locked loop (PLL) is a critical component in coherent communications receivers that is responsible for locking on to the carrier of a received modulated signal. A PLL adjusts the phase of a numerically-controlled oscillator to match that of the received signal. The top level entity of carrier recovery is shown in Fig. 4.24 where it consists of exact phase and received phase as input to calculate phase error then the error is fed to the loop path filter to generate the filtered error signal that control the numerically controlled oscillator as discussed in Chap. 1 [25, 26].

4.4.2.2 VHDL Simulation Results for DPSK Transceiver

VHDL simulation results for both 8DPSK, π/4DQPSK are shown in Fig. 4.25 and 4.26, respectively.

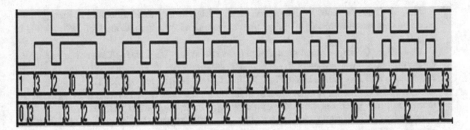

Fig. 4.25 VHDL simulation results for π/4DQPSK

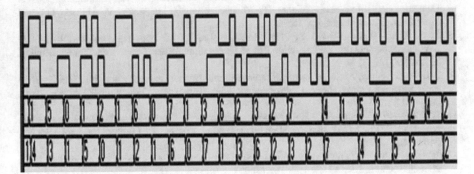

Fig. 4.26 VHDL simulation results for 8DPSK

4.5 Conclusion

In this chapter, DPSK (π/4DQPSK, 8DQPSK) modulation and demodulation are discussed. A SIMULINK and an HDL model is built. Timing and carrier recovery techniques are used to improve the BER performance.

References

1. H. Savla, *Design and Simulation of a Low Power Bluetooth Transceiver* (Master thesis, Sardar Patel University, India, Jan 2004)
2. H. Liu, Z. Sun, D. Tang, H. Huang, T. Kaneko, W. Deng, R. Wu, K. Okada, Matsuzawa. An ADPLL-centric Bluetooth low-energy transceiver with 2.3Mw interference-tolerant hybrid-loop receiver and 2.9mW single-point polar transmitter in 65nm CMOS. In *2018 IEEE International Solid – State Circuits Conference – (ISSCC)*. Feb. 2018, pp. 444–446. https://doi.org/10.1109/ISSCC.2018.8310375
3. M. Ding, X. Wang, P. Zhang, Y. He, S. Traferro, K. Shibata, M. Song, H. Korpela, K. Ueda, Y.H. Liu, C. Bachmann, K. Philips. A 0.8V 0.8mm² Bluetooth 5/BLE digital-intensive transceiver with a 2.3mW phase-tracking RX utilizing a hybrid loop Filter for interference resilience in 40nm CMOS. In *2018 IEEE International Solid – State Circuits Conference – (ISSCC)*. Feb. 2018, pp. 446–448. doi: https://doi.org/10.1109/ISSCC.2018.8310376

4. B. Hui, X. Tang, N. Gao, W. Zhang, X. Zhang, High order modulation format identification based on compressed sensing in optical fiber communication system. Chin. Opt. Lett. **14**, 14–18 (2016)

5. C. Li, J. Xiao, Q. Xu, A novel modulation classification for PSK and QAM signals in wireless communication. In *Proceedings of the IET International Conference on Communication Technology and Application (ICCTA)*, Beijing, China, 14–16 October 2011

6. A.G. Armada, L. Hanzo, A non-coherent multi-user large scale SIMO system relaying on M-ary DPSK. In *Proc. IEEE Int. Conf. Commun. (ICC)*, Jun. 2015, pp. 2517–2522

7. V.M. Baeza, A.G. Armada, Non-coherent massive SIMO system based on M-DPSK for Rician channels. IEEE Trans. Veh. Technol. **68**(3), 2413–2426 (2019)

8. J.B. Padhy, B. Patnaik, Optical wireless systems with DPSK and Manchester coding. Smart and Innovative Trends in Next Generation Computing Technologies, ser. Communications in Computer and Information Science, P. Bhattacharyya, H. G. Sastry, V. Marriboyina, and R. Sharma, Eds. Springer, pp. 155–167

9. L. Fu Xie, I. Wang-Hei Ho, S. Chang Liew, L. Lu, F.C.M. Lau, The feasibility of mobile physical-layer network coding with BPSK modulation. IEEE Trans. Veh. Technol. **66**(5), 3976–3990 (2017)

10. A.E.-S. El-Mahdy, Multiple tone interference of multicarrier frequency hopping BPSK system for a Rayleigh fading channel with channel estimation errors. Digit. Signal Process. **20**(3), 869–880 (2010)

11. K.H. Mun, S.M. Kang, S.K. Han, Multiple-noise-tolerant COOFDMA- PON uplink multiple access using AM-DAPSK-OFDM with reflective ONUs. J. Lightw. Technol. **36**(23), 5462–5469 (2018)

12. M. Ohm, J. Speidel, Quaternary optical ASK-DPSK and receivers with direct detection. IEEE Photon. Technol. Lett. **15**(1), 159–161 (2003)

13. M. Seimetz, M. Noelle, E. Patzak, Optical systems with high-order DPSK and star QAM modulation based on interferometric direct detection. J. Lightw. Technol. **25**(6), 1515–1530 (2007)

14. A. Gozzi, *Complexity and Performance Analysis of the Digital Modem of the Enhanced Data Rate Bluetooth Standard* (Pisa University, 2004)

15. B. Talha, M. Patzold, BEP analysis of M-ary PSK modulation schemes over double rice fading channels with EGC. In *Proc. 7th IEEE Int. Conf. Mobile Ad-Hoc Sensor Syst. (IEEE MASS)*, San Francisco, CA, USA, Nov. 2010, pp. 624–629

16. P. Supnithi, W. Wongtrairat, S. Tantaratana, Performance of M-PSK in mobile satellite communication over combined ionospheric scintillation and _at fading channels with MRC diversity. IEEE Trans. Wirel. Commun. **8**(7), 3360–3364 (2009)

17. Artech, *Digital Modulation Techniques* (ARTECH HOUSE, INC., 2000)

18. G.L. Stuber, *Principles of Mobile Communications* (Kluwer, Boston, 2011)

19. D. Tse, P. Viswanath, *Fundamentals of Wireless Communications* (Cambridge, 2005)

20. A. Molish, *Wireless Communications* (Wiley, 2011)

21. J. Decuir, Introducing Bluetooth smart: Part II: Applications and updates. IEEE Consum. Electron. Mag. **3**(2), 25–29 (2014)

22. A. Al Safi, B. Bazuin, *FPGA Based Implementation of BPSK and QPSK Modulators Using Address Reverse Accumulators* (IEEE UEMCON, 2016)

23. B.R. Jammu, H.K. Botcha, A.V. Sowjanya, N. Bodasingi, FPGA implementation of BASK-BFSK-BPSK-DPSK digital modulators using System Generator, ICCPCT, 2017

24. R. Reed, *Implementation of a BPSK Transceiver* (Bachelor of Science, University of Kansas, Lawrence, Kansas, 2004)

25. O.C. Ugweje, *Communication Systems Seminar* (The University of Akron, 2000)

26. S. Prot, K. Palmkvist, Baseband Modulation schemes, Electronic Systems, Dept. EE, LiTH, 2006

Chapter 5
Verification of the Integrated Bluetooth Modem

5.1 The Transceiver Block Diagram

In previous chapters, simulation and HDL implementation of both GFSK and DPSK (π/4DQPSK and 8DPSK) were discussed. Here we integrate both transceiver on a common hardware platform. The Bluetooth transceiver Block Diagram is shown in Fig. 5.1 [1–3].

At the transmitter side It can be noticed that different modulation schemes for Bluetooth Transceiver can be chosen using multiplexers ("00" GFSK, "01" π/4DQPSK, "10" 8DPSK), in case of GFSK the binary data is Gaussian filtered and then transmitted with carrier frequencies, in case of DPSK data is mapped to in-phase and quadrature values and then square root raised cosine filtered and finally transmitted by carrier frequency.

At the receiver side the same low path filters and the same ARCTAN function block are used for both GFSK and DPSK transceiver which saves hardware area that is why we chose this architecture as the receiver part for GFSK Transceiver as our target is low cost Bluetooth transceiver chip, but in case of GFSK the data is differentiated after the ARCTAN block [4–6]. The GFSK or DPSK are then sampled and enter the decision maker block, in case of DPSK they finally pass through serial to parallel converter. The carrier recovery and timing recovery blocks are included in the receiver side too [7–20].

5.2 Simulink Model for the Transceiver

The first stage of development of a wireless modem typically is a software simulation of the physical layer transmitter and receiver. The air interface is simulated with a channel model that tries to recreate real world conditions such as noise, fading, multipath, Doppler spread and path loss.

© The Author(s), under exclusive license to Springer Nature Switzerland AG 2022
K. S. Mohamed, *Bluetooth 5.0 Modem Design for IoT Devices*,
https://doi.org/10.1007/978-3-030-88626-4_5

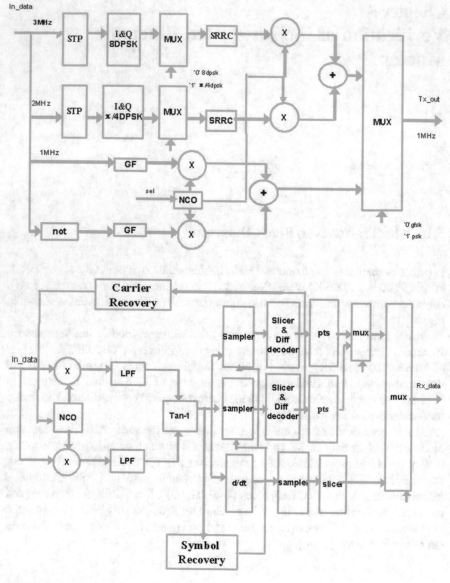

Fig. 5.1 The transceiver

Simulink is a block diagram environment for Model-Based Design. Its primary interface is a graphical block diagramming tool and a customizable set of block libraries. In The Simulink Model for the Transceiver, there are switches to run simulations with or without timing or carrier recovery and to choose between modulation techniques.

5.3 VHDL Model for the Transceiver

The Top level of the VHDL Model for The Bluetooth Transceiver is shown in Fig. 5.2 and its pin configuration is shown in Table 5.1.

5.3.1 Linear Feedback Shift Register

An LFSR is a shift register that, when clocked, advances the signal through the register from one bit to the next most-significant bit. Some of the outputs are combined in exclusive-OR configuration to form a feedback mechanism. A linear feedback shift register can be formed by performing exclusive-OR on the outputs of two or more of the flip-flops together and feeding those outputs back into the input of one of the flip-flops as shown in Fig. 5.3.

Linear feedback shift registers make extremely good pseudorandom pattern generators. When the outputs of the flip-flops are loaded with a seed value (anything

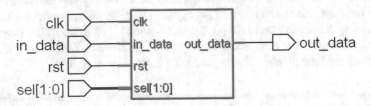

Fig. 5.2 VHDL model for Bluetooth transceiver

	Pin name	Description
Table 5.1 Pin configuration for top level of Bluetooth transceiver	Clk	Sampling frequency
	Rst	Enable high
	In_data	In_serial data
	sel	"10" GFSK "01" π/4DQPSK "00" 8DPSK
	out_data	The binary data after demodulation

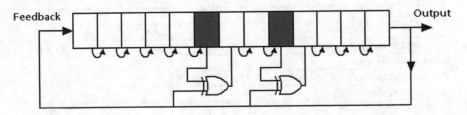

Fig. 5.3 LFSR schematic

except all 0 s, which would cause the LFSR to produce all 0 patterns) and when the LFSR is clocked, it will generate a pseudorandom pattern of 1 s and 0 s. Note that the only signal necessary to generate the test patterns is the clock. During all the simulations 30-Bit LFSR is used.

5.3.2 VHDL Simulation Results for the Transceiver

Figure 5.4 shows the VHDL simulation results for the transceiver where the data is randomly generated using 30-bit linear feedback shift register.

5.3.3 Synthesis Results of the Transceiver

One of the objects of this thesis to achieve synthesizable VHDL code for the Bluetooth transceiver, the design is synthesized using (FPGA advantage tool) and Mapped to Xilinx spartan3 FPGA (400pq208) kit, following successful RTL level simulation each unit was integrated and synthesized. Mapping report for the final system using Xilinx spartan3 (400pq 208) kit is shown in Table 5.2.The percentage of each component that contributes to the design is shown in Table 5.3. The rest of the 75% of the total number of 4 input LUTS is used as route-thro.

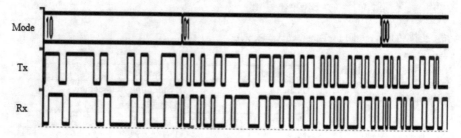

Fig. 5.4   VHDL simulation results for the transceiver

Table 5.2   Mapping result for the transceiver

Resource type	Used	Available	Utilization
Logic utilization			
Number of slice flip flops	2663	7168	37%
Total number of 4 input LUTS	4922	7168	68%
Number used as logic	3781		
Number used as route-thro	1211		
Number used as shift registers	8		
Number of mult 18 × 18	8	16	50%
Number of GCLKS	1	8	12%

Table 5.3 Utilization of each component in the transceiver

Component	Utilization	Total number	Total utilization
Numerically controlled oscillator	8%	2	16%
Numerically controlled clock	8%	1	8%
Arctan function block	6%	1	12%
Gaussian filter	1%	2	2%
SRRC filter	6%	2	12%
Low pass filter	6%	2	12%
Timing recovery	2.5%	1	2.5%
Carrier recovery	1.5%	1	1.5%
Others	2%	1	2%
Route-thro	32%		32%
			100%

Simulation

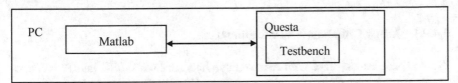

Hardware

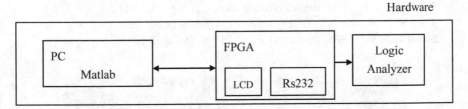

Fig. 5.5 Verification of system phases

5.4   Verification Strategy

The testing can be divided into two phases, a simulation phase where all testing is done on the PC and a hardware phase where testing is done on the hardware. In Fig. 5.5 an overview can be seen. Note that there are more steps involved in extracting test data in the hardware phase.

5.4.1   Simulation Verification

Questa tests were run when changes were made to the system; test benches were developed to read test input data produced by a MATLAB model and to generate MATLAB-readable output data. Data representation could then be changed to allow comparison with reference data produced by the MATLAB model and error patterns were studied to determine the cause of errors.

5.4.2   Hardware Verification

All the designs were thoroughly tested on the FPGA. Different hardware tests are discussed in the following sections.

5.4.2.1   Xilinx Chip-Scope Verification Test

We compare sent symbol with received symbols for both GFSK and PSK Transceiver using Xilinx Chip-Scope logic analyzer [8]. The Experiment Devices of the system are shown in Fig. 5.6.

The results for GFSK transceiver are shown in Figs. 5.7 and 5.8.

The results for π/4DQPSK transceiver are shown in Fig. 5.9.

The results for 8DPSK transceiver are shown in Fig. 5.10.

Fig. 5.6 Experiment devices of the system

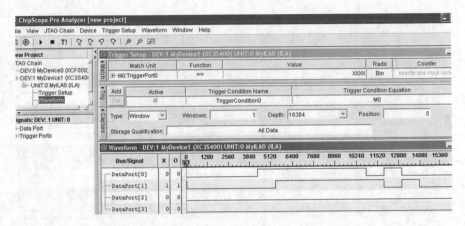

Fig. 5.7 GFSK verification result

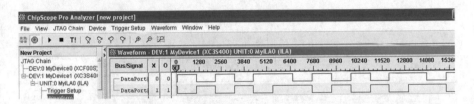

Fig. 5.8 GFSK verification result

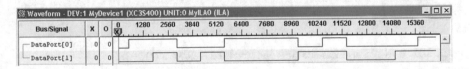

Fig. 5.9 π/4DQPSK verification result

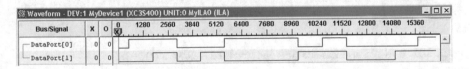

Fig. 5.10 8DPSK verification result

5.5   Tests Results

The tests have been passed successfully. The SIMULINK BER versus SNR curve and VHDL Measured/simulated E_b/N_0 versus SNR for Bluetooth Transceiver are shown in Figs. 5.11, 5.12 and 5.13; these curves are drawn at frequency offset = 150 KHz, phase offset = 0.1 RAD and timing offset = 0.1 us. Also Fig. 5.14 combines all the results.

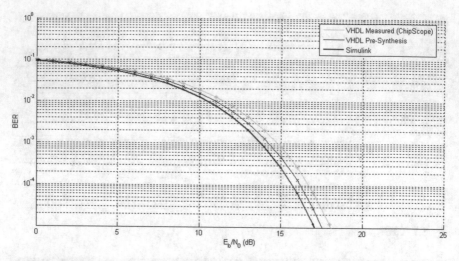

Fig. 5.11 E_b/N_0 versus measured BER and simulated BER for GFSK transceiver

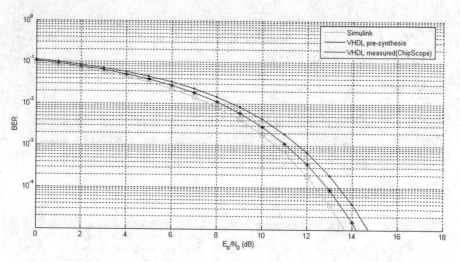

Fig. 5.12 E_b/N_0 versus measured BER and simulated BER for π/4DQPSK

5.6   Conclusion

In this chapter, integration of the transceiver is discussed, Simulink model for it was built and VHDL model too. System verification was done using both of simulation verification and hardware verifications, where all tests have been passed successfully.

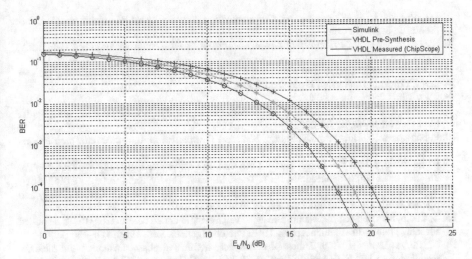

Fig. 5.13 E_b/N_0 versus measured BER and simulated BER for 8DPSK transceiver

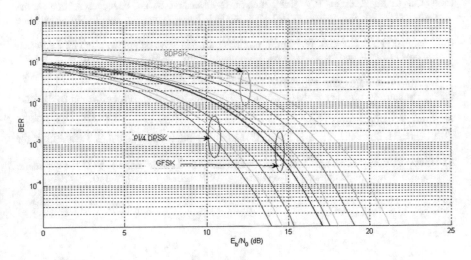

Fig. 5.14 E_b/N_0 Vs BER curves (GFSK & DPSK)

References

1. B. Xia, C. Xin, W. Sheng, A.Y. Valero-Lopez, E. Sanchez-Sinencio, A GFSK demodulator for low-IF Bluetooth receiver. IEEE J. Solid State Circuits **38**(8), 1397–1400 (2003)
2. J. Im, H. Kim, D.D. Wentzloff, A 335µW −72dBm receiver for FSK back-channel embedded in 5.8GHz Wi-Fi OFDM packets. In 2017 IEEE Radio Frequency Integrated Circuits Symposium (RFIC), 2017, pp. 176–179
3. K.S. Mohamed, *IP Cores Design from Specifications to Production: Modeling, Verification, Optimization, and Protection* (Springer, Cham, 2016)
4. M.C. Jeruchim et al., *Simulation of Communication Systems: Modeling, Methodology, and Techniques* (Kluwer, New York, 2000)

5. J.G. Proakis et al., *Contemporary Communication Systems Using MATLAB and Simulink* (Thomson & Books/Cole, USA, 2004)
6. W.H. Tranter et al., *Principles of Communication Systems Simulation* (Prentice Hall, USA, 2004)
7. S. Mukthavaram, *Design and FPGA Implementation of an Adaptive Demodulator* (Master thesis, Osmania University, Hyderabad, India, 1997)
8. www.xilinx.com
9. H. Savla, *Design and Simulation of a Low Power Bluetooth Transceiver* (Master thesis, Sardar Patel University, India, Jan 2004)
10. R. Schiphorst, F. Hoeksema, K. Slump, *Bluetooth Demodulation Algorithms and their Performance* (University of Twente, 2003)
11. HJ Bergveld, KMM van Kaam, DMW Leenaerts, A Low power highly digitized receiver for 2.4 GHz band GFSK applications. IEEE Trans. Microw. Theory Tech., 2005;53(2):453–461
12. A. Neubauer, A digital receiver architecture for bluetooth in 0.25 um CMOS technology and beyond. IEEE Trans. Circuits Syst. **54**(9), 2044–2053 (2007)
13. Y. Yadin, M. Orenstein, M. Shtaif, Balanced versus single-ended detection of DPSK: degraded advantage due to fiber nonlinearities. IEEE Photon. Technol. Lett. **19**(3), 164–166 (2007)
14. G. Xie, A. Dang, H. Guo, Effects of atmosphere dominated phase fluctuation and intensity scintillation to DPSK system. In Proc. IEEE Int. Conf. Commun. (ICC), Kyoto, Japan, June 2011, pp. 1–6
15. N. Chi, L. Xu, J. Zhang, P.V. Holm-Nielsen, C. Peucheret, S. Yu, P. Jeppesen, Improve the performance of orthogonal ASK/DPSK optical label switching by DC-balanced line encoding. J. Lightw. Technol. **24**(3), 1082–1092 (2006)
16. C.-H. Park, K.-S. Hong, S.-W. Nam, J.-H. Chang, Biased SNR estimation using pilot and data symbols in BPSK and QPSK systems. J. Commun. Netw. **16**(6), 583–591 (2014)
17. G. Hu, P. Zhao, Con_dence test for blind analysis of BPSK signals. IEEE Trans. Aerosp. Electron. Syst. **55**(2), 658–675 (2019)
18. P. Krishnan, Performance analysis of hybrid RF/FSO system using BPSK-SIM and DPSK-SIM over gamma-gamma turbulence channel with pointing errors for smart city applications. IEEE Access **6**, 75025–75032 (2018)
19. Y. Zhou, H. Zhang, D. Yuan, P. Zhang, H. Liu, Differential spatial modulation with BPSK for two-way relay wireless communications. IEEE Commun. Lett. **21**(6), 1361–1364 (2017)
20. H. Gu, Y. Wang, S. Hong, G. Gui, Blind channel identification aided generalized automatic modulation recognition based on deep learning. IEEE Access **7**, 110722–110729 (2019)

Chapter 6
Conclusions

Wireless sensor nodes in the Internet-of-Things (IoT) applications require ultralow-power consumption for radios to prolong battery lifetime. Bluetooth-low-energy (BLE) has been widely used in the IoT applications due to its ubiquitous deployment and easy access. Bluetooth 5 enhances BLE by providing a higher data rate and longer range.

This book has two major parts, Bluetooth signal processing and implementation, the application area of this book was a Bluetooth Transceiver. We consider that obtaining a model in SIMULINK which could serve as a test bench, was the most important purpose with the analysis of the communication system. This means that the underlying theory behind digital communication was treated lightly. The choice of algorithms was based on simulations in SIMULINK and "considerations on implementation": the CORDIC algorithm was favored over the ROM LUT approach to save the memory. The FDATOOL VHDL model was favored over the hand-crafted filter to save area. The synchronization algorithms improved the performance at the cost of the area. In the final architecture, we did the followings:

- Test under worst case SNR, time error and frequency offset.
- Integrate GFSK and DPSK data path.
- Retest whole system and optimization.
- An implementation of the Bluetooth Transceiver was completed and tested successfully on the FPGA.

Firstly, with regards of the modulation part, tests have been performed checking the output values of the filtering implementations through FDATOOL Moreover, simulation results show that the ISI and the BER value can be reduced further by increasing the length of the SRRC filter impulse response. As known, higher filter orders cost more in terms of chip area employed, therefore a cost/performance trade-off is required. The timing Recovery algorithm provides a direct estimation of the timing information and can be implemented with a digital circuit also the carrier recovery was implemented in digital domain. The whole system has been designed

K. S. Mohamed, *Bluetooth 5.0 Modem Design for IoT Devices*,
https://doi.org/10.1007/978-3-030-88626-4_6

and tested for achieving the desired data rate of 1-2-3 Mbps. Both the transmitter and the receiver have been simulated with an industrial simulator.

Moreover, Logic simulations along with pre-implementation timing analysis have been done to verify the proper working of the designed architecture. The simulated results were found to match with the desired performance criteria. Simulation results for the constituent blocks of the transmitter and receiver are discussed and shown along with the architectural diagram in the previous chapters.

FPGA flexibility and wide range of on-chip resources can thus yield very efficient implementations of adaptive receivers for current and future generations of wireless communications systems.

Index

Printed in the United States
by Baker & Taylor Publisher Services